Geol

Geol

by G. A. L. Johnson

Edited by
J. T. Greensmith

© THE GEOLOGISTS'
ASSOCIATION
1997

ISBN 0-900717-49-1

Geology of Hadrian's Wall

CONTENTS

PREFACE	(ii)
LIST OF FIGURES	(iv)
INTRODUCTION	1
LOGISTICS	3
HISTORY OF HADRIAN'S FRONTIER ZONE	6
CONSTRUCTION OF HADRIAN'S WALL	7
THE EASTERN SECTOR	17
THE CENTRAL SECTOR	36
THE WESTERN SECTOR	68
FURTHER READING	87

Geology of Hadrian's Wall

PREFACE

Rewarding time spent between 1948 and 1953 as an undergraduate and postgraduate student mapping and making a biostratigraphical survey along the central sector of Hadrian's Wall began a lasting interest in the connection between geology and archaeology. Nearly 50 years ago, the central sector of the Wall, between Chollerford and Greenhead, was still remote and isolated with public transport limited to the Tyne valley. The farming community rode ponies and geologists walked. For the scattered farms the monotony was broken by numerous travelling salesmen carrying every conceivable household need and by the occasional Methodist camp meeting at one of the farms on a Sunday afternoon. It was a friendly and hospitable community who gave open access to the countryside, except for a few sporting estates where the agent could be tricky if he came across you unexpectedly.

When it was suggested that an account of the geology of Hadrian's Wall be produced it seemed a pity to limit it to the central sector; an account of the geology of the Wall from sea to sea across the isthmus of northern England would be so much more satisfying. It has been no hardship to become better acquainted with the eastern and western sectors of the Wall where there are interesting sediments and antiquities even if the scenery is less spectacular.

The region traversed by Hadrian's Wall has been described by many writers and the present account should begin by acknowledging the debt we have to past and present scholars who have completed research in the area. The geology has been studied by a succession of workers from the middle of the 19th Century onwards. The line of the Wall is covered by 7 British Geological Survey maps, just four of which are supported by detailed descriptive memoirs. Broadly accepted stratigraphical terms have been used in the guide, not necessarily the most recent divisions. Line diagrams in the text (Figures 1, 13, 18, 21, 22, 27, 50, 54, 55 and 64) have boundaries derived from geological maps (sheets 13 Bellingham, 14 Morpeth, 15 Tynemouth, 17 Carlisle, 18 Brampton, 19 Hexham, 20 Newcastle and 21 Sunderland) at 1:50,000 or 1:63360 scales by permission of the Director, British Geological Survey; © NERC, all rights reserved.

Antiquarian interest in the Hadrian's Wall started in the 16th Century with a traverse of the Wall described in Camden's *Britannia* being the first detailed account. Later writers, particularly from the 18th Century onwards, have produced a succession of accounts of the Wall that contain valuable original commentary. Systematic archaeological investigations of Hadrian's Wall started in the early years of this century, with rapid development after the First World

Geology of Hadrian's Wall

War by active teams from the universities of Durham and Newcastle. Continuous progress over the past 50 years has been linked with increased visitor pressure and the provisions of public access and tourist facilities. English Heritage, National Trust, Vindolanda Trust, Northumberland and Cumbria County Councils, the Tyne and Wear Museums Service and other organisations are currently engaged in excavation and tourist activity.

The Hadrian's Wall path, a long distance National Trail along the whole length of the Wall, is at the development stage and will inevitably bring many new visitors to the already well trodden paths. There are management problems to be faced as the Wall become a major tourist attraction and high priority will have to be given to the protection and preservation of the Roman remains because they are irreplaceable. Similarly the scenery and environment along the Wall are fragile and will have to be managed carefully to retain isolation and control erosion, particularly where visitor pressure is high adjacent to popular stretches of the Wall in the central sector.

I am much indebted to many people who have advised and contributed to this account of the geology of Hadrian's Wall. It was Dr Eric Robinson who suggested that the account should be produced for the Geologists' Association Guides series and I am most grateful to him for his continued enthusiasm and for his careful reading of an early draft of the text. Dr Michael Jones also kindly read and advised on a draft of the text and supplied valuable contributions, particularly of the Haltwhistle Burn section (Figure 46) and Gilsland Spa, that are gratefully incorporated in the work. I am most indebted to Dr Trevor Greensmith, the Associations' Guides Editor, for his kindly supervision of the work and ever-ready advice and encouragement. Photographic illustrations come from the author's collection and were processed for publication by Gerry Dresser and Alan Carr. Field sketches were drawn for publication by Hazel and Pamela Johnson and the line diagrams were produced by computer based technology by Karen Atkinson. To all of these and others who have assisted in the production of this work, I tender most grateful thanks.

Geology of Hadrian's Wall

LIST OF FIGURES   **Page**

1. Geological map of northern England showing the 3 sectors of Hadrian's Wall   Facing 1
2. Section across Hadrian's frontier zone.   8
3. Reconstruction of a section of the Narrow Wall, near *Segedunum*, Wallsend.   8
4. Rock inscription of Legion XX *Valeria Victrix*, Wetheral.   9
5. Facing stones and core of the Wall at Poltross Burn milecastle 48.   9
6. Whin dolerite in the Wall core at Caw Gap.   9
7. Water-rounded cobbles in the core of the Wall near Greenhead.   10
8. Cawfields milecastle 42.   10
9. Banks East turret 52A.   11
10. Hadrian's Wall at Hare Hill, Banks.   11
11. Permian stones in the walls of Granary A5, *Arbeia Fort*, South Shields. 12
12. The ditch and berm at Black Carts, near Carrawburgh.   12
13. Geological map showing Hadrian's Wall in the eastern sector.   18-19
14. Reconstruction of the West Gate of *Arbeia* Fort.   20
15. Ruins of *Arbeia* Fort.   20
16. Geological section of Tynemouth Cliff.   21
17. Reconstructed SE corner of *Segedunum* Fort, Wallsend.   22
18. General succession of Westphalian, Coal Measures strata in the lower Tyne valley.   23
19. Hadrian's Broad Wall with turret 7B, Denton Burn, Newcastle upon Tyne.   25
20. Broad Wall at Heddon-on-the-Wall.   25
21. Geological map showing the Wall crossing Namurian strata in the eastern sector.   26-27
22. General succession of Namurian strata on the north side of the River Tyne.   29
23. Rock inscription at Written Crag.   31
24. The junction of Narrow Wall to Broad Wall at Planetrees, Brunton Bank.   31
25. Sandstone above the Great Limestone in Black Pastures Quarry, Brunton Bank.   32
26. Geological map of Hadrian's Wall in the central sector.   34-35
27. Asbian and Brigantian succession in the central sector.   37
28. The Great Limestone at Crindledykes, near Chesterholme.   39
29. Exposures in the River North Tyne, north of Chollerford.   40
30. View north from Limestone Corner, Carrawburgh.   41
31. The ditch before the Wall at Limestone Corner.   42
32. The Vallum cut through the Whin Sill at Limestone Corner.   42
33. Wedge holes cut by Roman workers, Limestone Corner.   43
34. The Mithraum at Carrawburgh.   45

Geology of Hadrian's Wall

35.	Escarpments north and south of Sewingshields Crags.	45
36.	Massive cross-bedded sandstone above the Lower Little Limestone at Queen's Crags.	47
37.	North-facing scarps of limestone, sandstone and Whin dolerite, south and east of Broomlee Lough.	48
38.	Housesteads Fort *(Vercovicium)*.	49
39.	*Vindolanda* Fort, Chesterholme.	50
40.	A Roman milestone, Stanegate Road, Chesterholme.	51
41.	The Whin Sill at Crag Lough.	52
42.	Castle Nick and Sycamore Gap in the Whin Sill escarpment.	52
43.	Peel Gap Tower.	53
44.	Green Slack glacial spillway, Winshields Crags.	54
45.	The Vallum running east from Cawfields Quarry.	55
46.	Brigantian Four Fathom Limestone near Cawfields.	56
47.	Geological map and sequence of strata at Haltwhistle Burn.	57
48.	Fellend Moss.	60
49.	Hadrian's Wall at turret 45A, Walltown.	60
50.	Geological map showing the course of the Wall over Asbian and Brigantian strata between Greenhead and Banks.	61
51.	Turret 48A at Gilsland.	64
52.	South gate and rampart at Birdoswald Fort.	66
53.	Pipers Sike turret 51A.	67
54.	Geological map showing the course of the Wall between Banks and Carlisle.	69
55.	General succession of Permian, Triassic and Jurassic strata of the Carlisle Basin, western sector of Hadrian's Wall.	70
56.	Ditch with line of Wall near Banks.	72
57.	Red sandstone Wall at Dovecote Bridge, Walton.	73
58.	Ditch and Wall line near turret 57B, Newtown.	74
59.	Remains of red sandstone Wall in hedge line at Newtown.	75
60.	Rock inscription at the Written Rock of Gelt.	75
61.	Ditch, berm and line of Wall at Oldwall.	76
62.	Ditch, berm and line of Wall at Birky Lane.	77
63.	Ditch crossing Brunstock Park, Tarraby.	77
64.	Geological map showing the course of Hadrian's Wall over drift-covered country between Carlisle and Bowness-on-Solway.	79
65.	St. Michael's Church, Burgh-by-Sands.	82
66.	The coastal Eastern Marsh near Watch Hill.	83
67.	Drumburgh Castle constructed of Roman stones.	83
68.	Coastal cliffs of boulder clay, Glasson.	84
69.	Eroding second terrace of marine alluvium on the Solway shore, west of Port Carlisle.	85
70.	The Solway Firth from the Banks, the north rampart of *Maia* Fort, Bowness-on-Solway.	85

Geology of Hadrian's Wall

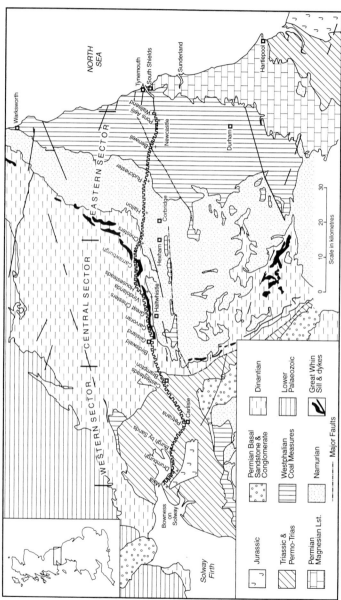

Figure 1. Geological map of northern England showing the three sectors of Hadrian's Wall.

Geology of Hadrian's Wall

INTRODUCTION

Hadrian's Wall crosses northern England from Wallsend, near the mouth of the River Tyne, in the east to Bowness-on-Solway, on the Solway Firth, in the west (Figure 1). It is 80 Roman miles long. A Roman mile is 1000 paces, approximately 1618 yds, so the length of the Wall is 73.5 statute miles or 118.25 km. The line of the Wall across the isthmus of northern England was carefully chosen at the shortest distance and took advantage of a natural barrier of high north-facing cliffs formed by the Great Whin Sill in the central sector (Figure 1). It was a frontier zone consisting of a stone or turf wall with a ditch or Vallum on the south side. It divided Roman Britain to the south from the barbarians to the north, prevented unwelcome intrusion and controlled the transit of people and animals across the frontier. The Wall does not seem to have had a special military purpose, but for much of the time the Romans were in Britain it formed the northern limit of their Empire. Today, it is one of the most celebrated Roman linear monuments and it has been designated a World Heritage Site.

The geology of the region of Hadrian's Wall divides naturally into three sectors. In the east, the course of the Wall for 41 km is over Upper Carboniferous, Coal Measures and 'Millstone Grit' (Namurian) strata with a low easterly dip. The succession is widely covered with glacial drift. For 37 km in the central sector, the Wall is underlain by Lower Carboniferous (Dinantian) sediments intruded by the quartz-dolerite Great Whin Sill. It is an upland region of ragged scarps with mainly well exposed strata. Here the regional dip veers from southeast to south and pronounced east to west running escarpments and dip slopes dominate the topography. The line of the Wall passes over the Permo-Trias-onto-Carboniferous unconformity just to the east of Brampton. In this western sector, the Wall crosses the New Red Sandstone of the Carlisle Basin for a further 40 km, but bedrock is often deeply buried in glacial drift. The Wall continues over this drift covered ground to the Solway Firth and the western termination at Bowness-on-Solway.

The north of England was repeatedly glaciated during the Quaternary when continental ice sheets spread over the region, probably on at least four occasions. Virtually all the known Quaternary sequence belongs to the last or Upper Devensian glaciation because deep erosion at this time removed the earlier Quaternary deposits. In the more lowland eastern and western sectors of the Wall almost continuous deep glacial till, sand and gravel mask the underlying bedrock. Only in the central sector, over higher ground, is bedrock well-exposed and in many places forms the foundation of the Wall.

In the central sector, striking scarp and dip-slope landforms arise from a combination of three factors. First, the Carboniferous succession, which ranges

Geology of Hadrian's Wall

from the Dinantian Jew Limestone to the Namurian Oakwood Limestone, is formed of Yoredale type cyclothems (Figures 22 & 27). This repeated or cyclic sequence of limestone, mudstone, sandstone, seatearth and coal produces an alternation of hard and soft beds of very different weathering characteristics. Into this succession was intruded, at the end of the Carboniferous, the quartz-dolerite Great Whin Sill, of varying thickness and often transgressing from one horizon to another. It is very durable and resistant to erosion in comparison to the enclosing sediments. Differential erosion of hard and softer dipping strata gives rise to escarpments. Second, after the deposition of the Carboniferous sediments, the Cheviot pluton in North Northumberland was uplifted resulting in doming of the sediments surrounding the igneous centre. This imparted a significant southerly dip to the strata in the central sector of the Roman Wall. Dips of around 15° south, coupled with the alternation of hard and soft beds, produced the celebrated E-W directed scarp and dip-slope topography. Third, in the central sector of the Wall, ice movement was from west to east, parallel to the strike of the beds. Glacial erratic boulders from the Lake District, including Shap Granite, have been recorded in the region. Movement of ice along the strike direction enhanced the scarp-dip-slope topography to the sharp relief now seen. If the movement of ice is across the strike it fills in the depressions with glacial debris, flattening out the topography, as in the eastern sector, east of Chollerford. Drilling in east Northumberland has revealed strong scarp and dip-slope topography in rockhead concealed under drift with a flat ground surface; the thickness of glacial deposits varies from nil to 80 m.

Variation in the the geological setting of the Wall caused changes in direction and in details of construction along its length. In the east sector, drift covered country with flat or rounded topography presented no difficulties to the builders and they planned the line of the Wall in long straight sections keeping to the high ground on the north side of the Tyne Valley. Building materials were abundant where bedrock sandstone comes near to the surface in drift-free ridges. The Vallum follows the Wall precisely with few deviations caused by topographic obstacles. In the central sector, the Wall follows the highest scarp formed by the Great Whin Sill for 16.5 km and is built on a secure quartz-dolerite foundation. Well exposed bedrock provided plenty of building materials. The Vallum lies at varying distances south of the Wall mainly dependent on geological constraints.

In the western sector, the Wall was built on deep drift overlying the New Red Sandstone of the Carlisle Basin and it tends to follow well drained boulder clay ridges. Bedrock sandstone only crops out in deep river valleys and is absent near to the Wall. Perhaps because of the lack of building stone and to speed up construction the Wall was initially built of turf in the western sector. It was replaced in stone at a later date when facing stones and the fill for the core had

Geology of Hadrian's Wall

to be carried long distances The Vallum does not follow the exact line of the Wall, but rather tends to cut off corners and take shortest routes.

Two accounts have been published on the geology of the Wall, John Hodgson (1822) gave an account of the central sector with a systematic study of the quarries and George Tate (1867) produced a fuller description of the course of the Wall as an appendix to the enlarged third edition of "The Roman Wall" by J. Collingwood Bruce. Geological Survey memoirs of the one-inch series sheets 11, 16 and 17 (Longtown, Silloth and Carlisle), 18 (Brampton), 13 (Bellingham) and 21 (Sunderland) give much detail of the geology of the region about the Wall. More general accounts of the geology of northern England have been produced by Taylor *et al.* (1971) and Johnson (1995). These and other publications will be mentioned in the following pages and in the list of Further Reading (p. 87).

LOGISTICS

You can reach Hadrian's Wall by many routes. From the south and east via the A1, A68 or A69; from the south and west via the A6, A686 or A69; from Scotland by the A74 or A68. Of these routes the A68 is recommended because it is the ancient Roman Road of Dere Street that leads to Corbridge, the Roman fort and settlement of *Corstopitum (Coriosopitum)*, 4 km south of the Wall. By car and cycle the Military Road (B6318) provides easy access to the best preserved and most scenic central sector of the Wall between Heddon-on-the-Wall and Gilsland. Using public transport, the Tyne Valley rail line runs almost parallel to Hadrian's Wall on the south and connects with Intercity trains at Newcastle and Carlisle, and Newcastle airport. It takes about one hour from Carlisle to Hexham and about 40 minutes from Newcastle to Hexham. There is also a regular bus service on the A69 from Newcastle to Carlisle that stops at villages in the Tyne Valley south of the Wall, but the Wall is often 5 km distant uphill on the north side of the valley. During holiday periods, a Hadrian's Wall bus service runs on the B6318 from Hexham to Haltwhistle calling at the many Roman sites and connecting with rail and coach services. Using this bus, car drivers can enjoy a walk along the Wall and return to the car park by bus.

The range of accommodation is excellent with many hotels and guest houses to the south of the Wall. There are five Youth Hostels at Newcastle, Acomb (Hexham), Once Brewed (Bardon Mill), Greenhead and Carlisle. The three central hostels command the best walking, geology and Roman remains and are very popular. Further information on access, accommodation and guide books can be obtained from either the Hexham Tourist Information Centre, Hallgate, Hexham, Northumberland NE46 1XD (Tel. 01434-605225) or The Northumberland National Park Visitor Centre, Military Road, Bardon Mill, Hexham, Northumberland, NE47 7AN (Tel. 01434-344396).

Geology of Hadrian's Wall

The climate in the much-visited central sector of the Wall is similar to other uplands in northern England. The Wall reaches 375 m above sea-level on Winshields Crags and crosses high ground for more than 20 km. It can be sunny and still on hot summer days, but more usually it is windy and unsettled and there is a high rainfall. During the winter months it is cold and wet and there can be some snow cover between October and April. Windproof and waterproof clothing and an extra sweater are advisable on most days and good footwear is needed because the walking is rough. The Hadrian's Wall Long Distance footpath runs close to the Wall and this and other footpaths are well signposted. Off the footpaths, most of the land is privately owned and permission for access to geological sites is necessary. Geologists visiting the area are reminded that the Geologists' Association 'Geological Fieldwork Code' gives sound advice on getting the most out of geology in the field and keeping out of trouble. Always regard access to geological sites as a privilege to be respected by exemplary conduct.

Maps of the Wall are covered by Ordnance Survey, 1:50,000 scale Landranger Series sheets 85, 86, 87 and 88. On the larger scale, 1:25,000 Pathfinder Series, sheets 534, 544-550, 557-558 are appropriate. The geology of the region through which the Wall passes is covered by the following 1:50,000 scale British Geological Survey sheets: 13 (Bellingham), 14 (Morpeth), 15 (Tynemouth), 17 (Carlisle), 18 (Brampton), 19 (Hexham), 20 (Newcastle) and 21 (Sunderland). A most useful additional map is the Ordnance Survey historical map and guide of Hadrian's Wall which gives topography and communications with details of the Roman remains. It is particularly valuable, since it distinguishes extant and visible Roman monuments from those where the site is not excavated and little can be seen. Local guide books are on sale at the major Roman sites along the Wall and longer accounts of the Roman occupation of north Britain include *Hadrian's Wall* by Breeze and Dobson (1978) and two new books *Hadrian's Wall* by Stephen Johnson (1989) and *Housesteads* by James Crow (1995), both are *English Heritage* Series books and are recommended. Routes and itineraries for walking along the line of Hadrian's Wall from east to west are described in two new guides. *The Wall Walk* by Mark Richards (1993) gives detailed directions for finding footpaths on or near to the Wall with many maps and line diagrams. *Walking the Wall* by Tony Hopkins (1993) describes the line of the Wall and comments more broadly on the scenery, flora and fauna to be seen along the route; it is again beautifully illustrated with field sketches.

Geology of Hadrian's Wall

Table 1. Outline chronology of the Roman Frontier in Britain

Dates AD		
410	–	End of Roman Britain.
407	–	Imperial authority and payment of the Wall garrison ceased.
300-400	–	Rare artefacts show that Roman occupation of the Wall continued in the 4th Century; gradual loss of contact with Rome.
383	–	Magnus Maximus, governor of Britain, took the main part of the Wall garrison to Gaul on campaign and was defeated.
360-367	–	Barbarian revolt put down by Count Theodosius. Renovation of forts on the Wall follow.
305-306	–	Victory over the Picts by Emperor Constantinus.
300	–	Concerted raids by the Picts.
296-305	–	Last building inscription placed on Hadrian's Wall at Birdoswald.
210-350	–	Static frontier; the Wall garrison remained in place with little change despite the legions being withdrawn from Britain.
208-209	–	Emperor Severus' campaign in Britain to restore order pressed north to Aberdeen; renovation of the Wall continues.
c197	–	Governor Virius Lupus forced to buy peace from the barbarians.
c160-200	–	Repair and rebuilding of the Wall and milecastles; some turrets demolished; *Arbeia* (South Shields) developed as a supply base for the Roman Army garrison in north Britain; the turf wall from the Irthing west to Bowness rebuilt in stone.
c163	–	Antonine Wall abandoned, Scottish frontier given up finally.
c162	–	Brief return to the Antonine Wall frontier.
158	–	Repairs and rebuilding of parts of Hadrian's Wall by Legion VI.
c157-158	–	Antonine Wall abandoned, return to Hadrian's Wall frontier by the end of the 150's.
c150	–	Peak of Antonine advance into Scotland reaching the River Tay.
c145	–	Antonine turf wall across the isthmus from Forth to Clyde completed with 26 forts and fortlets along its length.
c140-145	–	Hadrian's Wall Vallum partly backfilled and causeways formed.
c140	–	Hadrian's Wall milecastles and turrets abandoned, but the forts continued to be manned by the legions.
139-140	–	New advance into southern Scotland.
138	–	Death of Hadrian, Antoninus Pius becomes Emperor.
c130	–	Hadrian's Wall forts, milecastles and turrets with uncertain lengths of curtain wall completed from Tyne to Solway.
c124-125	–	Forts moved up to the line of the Wall, change from broad gauge to narrow gauge wall ordered; turf wall construction started from the River Irthing to the Solway.
c122	–	Work on the broad gauge stone wall from Newcastle to the North Tyne probably started by the legions II *Augusta*, XX *Valeria Victrix* and VI *Victrix*.
122	–	Hadrian visited Britain and ordered the construction of a massive frontier wall across the Tyne to Solway isthmus on the north side of the Stanegate Road; Nepos made responsible for building work.
121	–	Hadrian began a tour of the frontiers of the empire and appointed his friend Aulus Platorius Nepos to be governor of Britain.
117	–	Hadrian became emperor.
105	–	Stanegate Road forts harassed by northern tribes.
97	–	Trajan became emperor and later proposed a frontier wall to protect the forts on the Stanegate Road.
c80's-90's	–	Construction of the Stanegate Road between Corbridge and Carlisle. Forts built initially 21 km apart and later at 10 km apart.
c87-105	–	Withdrawal of the Legion II *Adiutrix* from Britain to the Danube owing to a Dacian invasion; withdrawal from forts in Scotland and north of the Newcastle - Carlisle line including Corbridge.
78-85	–	Agricola made governor; campaigns in north Britain and Scotland.
c71-72	–	Defeat of the northern tribes, including Brigantes, at Stanwick.
60	–	Boudicca's revolt, defeat of the southern tribes.
A.D. 43	–	Emperor Claudius invaded Britain decisively with four legions.
55-54 B.C.	–	Julius Caesar raided Britain and defeated the tribes of Kent.

Geology of Hadrian's Wall

HISTORY OF HADRIAN'S FRONTIER ZONE

Roman forces under Emperor Claudius arrived in Britain in A.D. 43 and gradually spread northwards by conquest and subsequent Romanisation of the British tribes (Table 1). By A.D. 70 they had established bases at York and Chester, to the north of which lay the territory of the Brigantes - a client territory. In A.D. 71 relations with the Brigantes broke down and after a battle at Stanwick, near Scotch Corner, the Romans spread north to Northumberland and south Scotland. In A.D. 78 under Agricola, they pushed further into Scotland and established roads and fortlets as far north as Perth. Two main routes were constructed, one up the west side of the country through Chester and Carlisle to Scotland and the other up the east side from York to Corbridge to Rochester (Carter Bar) and Scotland; this is Dere Street (A68).

Under Agricola, a fort was established at Corbridge and a road constructed west, along the Tyne valley, to link Dere Street with the western route at Carlisle; this road was called Stanegate. A series of forts were built along Stanegate for its defence, at Newbrough, *Vindolanda, Carvoran,* Throp, Nether Denton and Brampton Old Church with a smaller fortlet at Haltwhistle Burn. No wall existed at this time and the Romans, in the Stanegate forts, were harrassed and attacked particularly about A.D. 105.

In A.D. 117 Hadrian became emperor replacing Trajan who had suggested a wall across England to define the limits of the Roman Empire and to defend the Stanegate frontier road. Hadrian set out on a tour of his empire in A.D. 121 and during a visit to Britain ordered the construction of the wall proposed by Trajan. The frontier defences of ditch, Wall and Vallum was the idea of Hadrian and his friend Aulus Platorius Nepos, legate in Britain, who was responsible for putting the scheme into effect (Figure 2). It was planned to run north of Stanegate on high ground, much of it, in the central sector, along the high scarp of the Great Whin Sill (Figure 1).

Inscriptions found along the line of the Wall show that it was constructed by soldiers from the three legions of Britain, the II *Augusta* from Caerleon, South Wales, the XX *Valeria Victrix* from Chester and the VI *Victrix* from York. Each legion was made up of about 5000 men (Roman citizens) divided into centuries. They were infantrymen with some cavalrymen and were well armed, highly trained and disciplined. Each legion had its engineering corps, building and maintenance staff who were used for major building programmes. Auxilliary units raised from the tribes of the Empire were another branch of the Roman army. They were not Roman citizens and took no part in building Hadrian's Wall except perhaps fetching and carrying. During building, the legions were responsible for lengths of 8 or 9 km and minor constructional differences in

Geology of Hadrian's Wall

milecastles, turrets and the curtain wall indicate work done by an individual legion, but which legion is often uncertain.

Building the wall started in A.D. 122 and it seems to have been mainly completed by A.D. 130, about 8 years after commencement, but as the political and military situation changed, alterations to forts, gateways and ditches continued to about A.D. 160. Hadrian's successor Antoninus (A.D. 138) moved the frontier forward again and established the Antonine Wall of turf between the Forth and Clyde estuaries. At this peak of Roman advance, fortlets were constructed as far north as the Tay, but all were abandoned in A.D. 157-158. Subsequently, there were troubles in the north and after A.D. 200 the Emperor Severus had to restore order and make major alterations and repairs to Hadrian's Wall. Some milecastles and turrets were restructured and the Vallum was cleaned out with a new marginal mound of debris formed on the south lip (Figure 2). The Roman frontier was garrisoned until the end of the fourth century when the military failed to deal adequately with rebellion and invasion from overseas. During this unrest, centralized Roman administration lapsed and was replaced by smaller regional chiefdoms that may have been in part the ancient British tribal territories. The Wall, no longer necessary, ceased to be manned in about A.D. 410.

During the centuries after the Roman occupation, the Wall and its associated settlements have been plundered by local people seeking the dressed stone as an easy source of building material. The greatest destruction took place after the 1745 Jacobite Rebellion when the Scots, under Bonny Prince Charlie, made easy passage down the west side of England while the English Army, stationed at Newcastle on the east coast, was unable to cross the country. After this campaign the English Army, under General Wade, decided to build a new east-west road from Newcastle to Carlisle to speed troop movements. Despite protests from local antiquarians. the new road, still known as the Military Road (B6318), was built on top of the foundations of the Wall from milecastle 4 in Newcastle to milecastle 33 near Sewingshields, 48 km to the west. Only the path of the Wall in the central sector over high and isolated crags, between Sewingshields and Greenhead, saved it from being almost totally concealed. In retrospect, by burying the Wall below the road, its course and foundations have been preserved safely to the present day using the most practical means to achieve this.

CONSTRUCTION OF HADRIAN'S WALL

The Wall was planned by Hadrian and Platorius Nepos to be built progressively from east to west. It was to be of stone, 3 m wide and up to 4.6 m high with possibly 1.8 m of crenellated parapet (Figures 2 & 3).

Geology of Hadrian's Wall

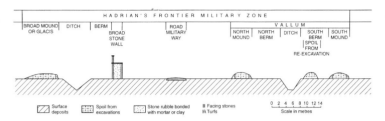

Figure 2. Section across Hadrian's frontier zone showing the relative positions of the Ditch, Wall, Military Way and Vallum.

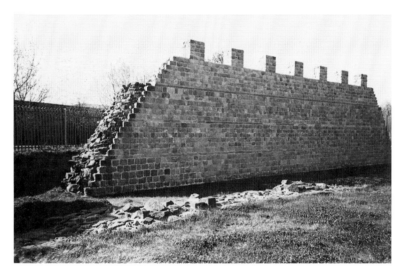

Figure 3. Reconstruction of a section of the Narrow Wall near Segedunum, Wallsend. The original Wall foundations are in front of the reconstruction.

The foundation was shallow and formed of rough flagstones, cobbles and puddled clay. In the eastern and central sectors the facing stones were cut from durable Carboniferous sandstone that quickly weathers to a sombre grey colour. The abundance of exposures over the Carboniferous outcrop allowed stone to be obtained locally from many small quarries; there are few large sandstone quarries even today. Some Roman quarries are identified by inscriptions left by the quarrymen, but as would be expected, these tend to date from the later periods of repairs and reconstruction of the Wall, earlier inscriptions being lost by later quarrying. A Roman quarrry inscription by Legion XX *Valeria Victrix*

Geology of Hadrian's Wall

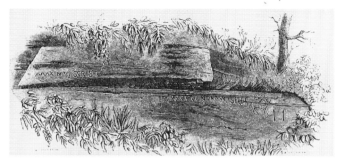

Figure 4. *Rock inscription of Legion XX Valeria Victrix in red Triassic sandstone, gorge of the River Eden at Wetheral (Burce, 1867, p.290).*

Figure 5. *Facing stones and core of the Wall at Poltross Burn milecastle 48. Rounded cobbles in the core come from the bed of the burn.*

Figure 6. *Whin dolerite in the Wall core at Caw Gap west of Winshields Crags. Ancient quarries in the background are in the Whin Sill and may have been opened by the Romans.*

Geology of Hadrian's Wall

Figure 7. Water-rounded cobbles from the Tipalt Burn in the core of the Wall where it crosses the B6318 road near Greenhead.

Figure 8. Cawfields milecastle 42 built on the dip-slope of the Whin Sill.

Geology of Hadrian's Wall

Figure 9. Banks East turret 52A, a Turf Wall turret with adjoining later Narrow Wall in stone.

Figure 10. Hadrian's Wall at Hare Hill, Banks. The highest part of the original Wall to have survived. The second stone from the left on the ninth course from the top is a centurial stone and has the initals 'PP' for 'primus pilus' the senior centurion of the cohort building the Wall.

Geology of Hadrian's Wall

Figure 11. Permian Concretionary Limestone building stones in the walls of Granary A5, Arbeia Fort, South Shields.

Figure 12. The ditch and berm on the north of Hadrian's Wall (seen in the background) at Black Carts, Limestone Bank, near Carrawburgh.

Geology of Hadrian's Wall

in red Triassic sandstone from Wetheral is an example (Figure 4). In preference, well-cemented, massive, medium-grained sandstone was used for the facing stones. They were cut to manageable size, often 150 mm high, 220 mm wide and 400 mm long running into the core of the Wall, but larger stones were also used particularly in the bottom courses and gateways of the Wall.

Between the facing stones on both sides of the Wall, the core was formed of rough stones bonded with puddled clay or in some places mortar (Figure 5). The fill of the core is dominated by quarried, angular sandstone slabs of local origin with few anomalous far-travelled stones perhaps derived from glacial boulder clay. At Denton Burn, just east of turret 7B, there are several pieces of Carboniferous dark grey crinoidal limestone in the core among many slabs of Coal Measures sandstone. The limestone could be derived from boulder clay or it might have been brought into the area from the east or north for burning to make lime. On the Whin Sill outcrop between Sewingshields and Greenhead, much of the fill of the Wall is angular slabs of quartz-dolerite produced in small quarries near the Wall (Figure 6). This is well seen at Sewingshields Crags, Peel Crags, Caw Gap and Cawfields. Near rivers and streams, where rounded cobbles and boulders are abundant, this material is incorporated into the Wall core; such as near Greenhead by the B6318 road at NY 657661 where the core is composed of rounded stones probably from nearby Tipalt Burn (Figure 7).

Every Roman mile, a milecastle was build into the Wall; 80 of them in all. Each milecastle would house about 30 soldiers and incorporated a gateway through the wall (Figure 8). In between the milecastles were two turrets, each a third of a mile apart, providing shelter for 4 men (Figure 9). By convention the milecastles and turrets are numbered from east to west. Milecastle 1 is one mile west of the termination of the Wall against the River Tyne at Wallsend and milecastle 80 is beside the Solway Firth. Turrets are numbered according to the adjacent milecastles thus: milecastle 1, turret 1A and 1B; milecastle 2, turret 2A and 2B; milecastle 3, etc. The milecastles and turrets are shown and numbered on the Ordnance Survey maps and form valuable reference points along the Wall.

When work on the Wall started in A.D. 122 it was planned to be 3 m wide, but by A.D. 124 it was decided to reduce the width to 2.3 m to save materials and speed construction; intermediate dimensions are also found. At this stage it was realized that the Stanegate forts were too distant from the Wall and 16 new forts were planned and constructed at 5 Roman mile intervals along the line of the Wall (Figure 1). These forts were major gateways, trading posts, lookouts and barracks for the garrison that manned the Wall.

In the western sector, over the Carlisle basin, the stone Wall was built of red Permo-Triassic Sandstone. This is a less durable stone than the Carboniferous

Geology of Hadrian's Wall

sandstone and almost all of it has weathered away or been recycled as building stone. One short length of red sandstone Wall was visible until 1983 at Dovecote Bridge, east of Walltown (NY 527644), but this was deteriorating so quickly owing to natural weathering that it has been covered with clay soil; it is now an elongate grassy mound. At the time of their geological survey of the Brampton area in the 1920s, Trotter and Hollingworth (1932) mention that the ruins of the Roman Wall are confined to the Carboniferous outcrop and the facing stones are invariably Carboniferous sandstone. In the region west of Carlisle, Dixon *et al.* (1926) report that, in the absence of rock suitable for building, the stones of the Roman Wall have been extensively used and the almost complete disappearance of the Wall is accounted for. They note that the churches of Burgh-by-Sands and Great Orton and the castle of Drumburgh are largely constructed of Roman stones.

At Hare Hill west of Banks (NY 564646), near to the junction of the Carboniferous and Permo-Trias, a short section of Wall has 17 courses of facing stones on the north side composed of grey Carboniferous sandstone and red-brown Triassic Sherwood sandstone (Figure 10). The core contains abundant angular grey-and red sandstone blocks with Carboniferous limestone and rounded boulders of dolerite and lava that might have come from the alluvium of the River Irthing. Rounded boulders of igneous rock, including coarse-grained porphyritic rocks, are also present in the core of the Wall west of Birdoswald and slabs of limestone occur in the core at Banks East turret 52A.

Use of locally available building material is well-shown at *Arbeia*, the supply base fort at South Shields. Here white, spheroidal, Permian Concretionary Limestone was used for building stones in the 2nd century double granary A5 near the west gate (Figure 11). In other buildings mixtures of sandstone and limestone were used including large limestone blocks in the lower courses of the north gateway.

The core of the Wall may be set in mortar rather than clay and for this large quantities of limestone were needed for lime burning. In the east, the Permian Magnesian Limestone of the Durham coast is the only source of limestone; there is a history of lime burning at Fulwell, north of Sunderland from early times up to the 1960's and it was an important local industry. Possibly, lime from the Magnesian Limestone was used to make mortar in the eastern sector of the Wall over the Coal Measures where limestone is absent. The thin limestones of the Namurian succession west of Heddon-on-the-Wall have been burnt for lime, particularly the Corbridge and Thornbrough limestones north of Corbridge where there are extensive old quarries. Lime mortar for building the Wall, fort and settlement of *Corstopitum (Coriosopitum)* may have come from this source. Westwards, thick limestone bands at the base of the Namurian and in the

Geology of Hadrian's Wall

underlying Brigantian enter near the River North Tyne and between here and Greenhead there is much good limestone that was burnt for lime in small stone kilns up to the 1950's. Variable quantities of clay and silt in these limestones produce a natural cement when burnt in the kiln. West of Greenhead, limestone is present in thinner bands in the Carboniferous sequence as far as Banks and Lanercost and it was probably burnt for lime by the builders of the Wall. Angular fragments of bioclastic limestone are present in the Wall core at Banks East turret 52A and Hare Hill, Banks suggesting that the local limestone bands were being worked in Roman times.

An earth or turf wall was built between the River Irthing and the Solway Firth during the early stage of development of the frontier. It is believed to have been built by two of the legions that were drafted to construct the stone wall westwards from Newcastle. This probably took place during the time of change, when the forts were moved forward to the line of the Wall and the width of the Wall was reduced to speed progress. There was new impetus to complete the Wall right across the isthmus of northern England. Owing to the deep glacial drift cover west of Greenhead, the places where building stone could be obtained was limited. An earth and turf wall had the advantage that no stone was needed and no mortar. The raw material was abundant, it was quick to build and could be replaced by stone when necessary. The importance of the lack of outcrops of sandstone and limestone near to the line of the Wall in the west is uncertain because the whole length, from Wallsend to the Solway, stood in stone in the 3rd Century.

The Turf Wall was 6 m wide at the base and perhaps 4.3 m high. It was constructed of materials available on site, turfs, earth, peaty silt etc. on a foundation of coursed turf 3 or 4 layers thick. The front was nearly vertical and the back rose at an angle of about 75°. As with the stone wall, the form of the top of the turf wall is uncertain. A massive stone wall is regarded as normal for the Roman frontier in Britain, but compared with other Roman frontiers, it is the stone wall that is unusual not the turf wall. Hadrian's stone wall is unique in its width and massiveness. On other Roman frontiers, narrow stone walls and dry stone walls were constructed, wooden fences were used or barriers were formed of turf.

The frontier zone of Hadrian and Nepos was completed by a ditch before the Wall and a further ditch system on the south side (Figure 2). In front of the Wall the ditch was 9.2 m wide and 3 m deep excavated 6 m from the north face of the Wall except where it was constructed above precipitous cliffs (Figure 12). On the south side of the Wall there was a marching road or military way about 4.8 m wide with cambered surface metalled with broken stone. Beyond this was the Vallum, a flat bottomed ditch 6 m wide and 3 m deep, which marked the southern

Geology of Hadrian's Wall

limit of the military zone. Spoil from excavating the Vallum was piled in mounds 1.8 m high and 6 m wide at a distance of 9.2 m on either side. The ditches and mounds are often clearly visible even when no signs of the Wall can be seen. It is noteworthy that the Romans used the word Vallum for the whole frontier zone of Wall and ditches, but the present day restricted use of the word for the ditch system south of the Wall also comes from antiquity.

Geology of Hadrian's Wall

THE EASTERN SECTOR

Hadrian's Wall starts against the north bank of the River Tyne at Wallsend, about 7 km west of the North Sea coast (Figure 13). The mouth of the Tyne was guarded by two Roman stations on Tynemouth Cliff to the north and at South Shields on the south. The small promontory at Tynemouth Cliff has been almost continuously occupied since Roman times and it is now covered by the remains of a priory with medieval fortifications. Roman stones with inscriptions have been recovered here, but there are no visible remains. South Shields has its origins in the Roman Fort of *Arbeia* beside a harbour at the mouth of the River Tyne where sea-going ships brought in supplies for Hadrian's Wall. It is a large 3rd Century fort with extensive store houses as would be expected at a supply base. The fort is being further excavated each year and a significant part of it has now been exposed. There is a museum at the entrance in Baring Street, Lawe Top. A full size reconstruction of the west gate is a major attraction and there is much to see (Figures 14 & 15).

On the North Sea coast near the mouth of the Tyne, Permian and Carboniferous rocks are well-exposed in accessible cliffs. The Permian forms outliers at Tynemouth Cliff and Cullercoats to the north of the Tyne where the basal Yellow Sands, Marl Slate and Magnesian Limestone are exposed. Tynemouth Cliff (Figure 16) is particularly instructive and shows the Permian sequence unconformably overlying reddened Middle Coal Measures (Jones, 1967; Land, 1974). Access to the section is by the North Pier, Tynemouth with descent to the beach by steps beside Tynemouth Cliff. Visits should be planned for low tide when the celebrated Tertiary tholeiite (a variety of basalt) Tynemouth Dyke is conspicuous running close and almost parallel to the North Pier (Teall, 1884). The Middle Coal Measures form almost continuous cliffs for 8 km north from Tynemouth to Seaton Sluice and this is the best section of these rocks in the region (Jones, 1967; Land, 1974); access to the cliffs at Tynemouth, Cullercoats, St. Mary's Island, Hartley and Seaton Sluice is best at low tide. For 38 km to the south of the Tyne, the Permian succession is exposed in sea cliffs at Marsden, Roker, Hendon, Seaham and Blackhall Rocks. They provide excellent sections through the full sequence to the Trias including the celebrated Magnesian Limestone barrier reef complex (Smith, 1994, 1985).

WALLSEND. The fort at the eastern terminus of Hadrian's Wall was *Segedunum* at Buddle Street, Wallsend (NZ 300660). The Narrow Wall connected the fort with the fort at Newcastle *(Pons Aelius),* so this eastern extension was a later addition to the original curtain wall. *Segedunum* is on the north bank of the River Tyne and its purpose was to prevent unauthorised crossings of the river at low water. An extension of the Wall ran from the SE corner of the fort to the edge of the river a distance of about 400 m; the foundations have been reconstructed in Roman stones beside the fort (Figure 17).

Geology of Hadrian's Wall

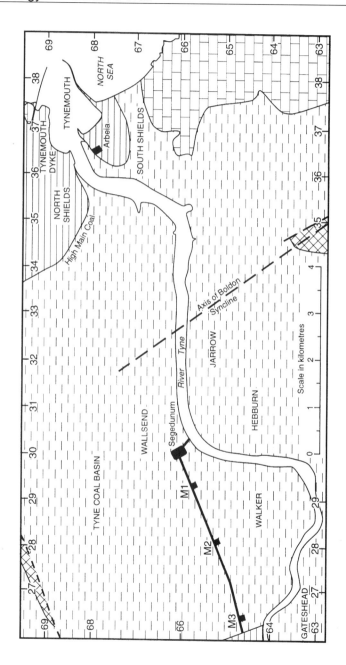

Geology of Hadrian's Wall

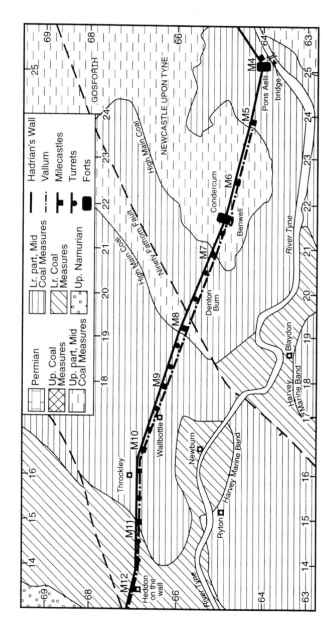

Figure 13. Geological map showing Hadrian's Wall crossing Westphalian Coal Measures in the eastern sector.

Geology of Hadrian's Wall

Figure 14. Reconstruction of the West Gate of Arbeia *Fort, South Shields.*

Figure 15. Ruins of Arbeia *Fort with reconstructed West Gate in the background.*

Geology of Hadrian's Wall

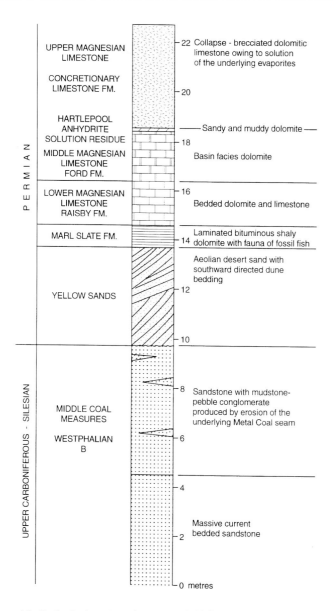

Figure 16. Geological section of Tynemouth Cliff showing the Permian unconformity.

Geology of Hadrian's Wall

Figure 17. Reconstructed SE corner of Segedunum *Fort, Wallsend, with the rebuilt foundation of the Wall leading down to the River Tyne. The river and shipyards in the background.*

The fort has been excavated and it is now being re-exposed and consolidated for visitors to see; only the headquarters building is visible at the time of writing. A small section of Narrow Wall footings can be seen on the north side of Buddle Street leading away from the fort's west gate towards Newcastle and a full size reconstruction of the Narrow Wall has been built here recently (Figure 3). Between the fort and the section of Narrow Wall, B Pit of Wallsend Colliery was sunk at the end of the 18th Century; the colliery closed in 1855. The capped shaft and complex of pit-head buildings are currently being excavated. The fort, Wall and colliery are to be preserved in a new park in Wallsend that is planned to be completed in Spring 2000 by the Tyne and Wear Museums.
Quaternary till covers the Carboniferous bedrock over much of the eastern sector of the Wall. The Coal Measures (Figure 18) can be seen in small exposures and particularly in the steep north bank of the River Tyne at Walker Riverside Park, St. Antony's (NZ 285630). Here, beds high in the Middle Coal Measures between the Ryhope Five-Quarter Coal and the Hylton Marine Band are exposed (Smith, 1994, p. 40). Thin drift in this region allowed access to thick sandstones for building the Wall, but the quarries have disappeared long ago.

Geology of Hadrian's Wall

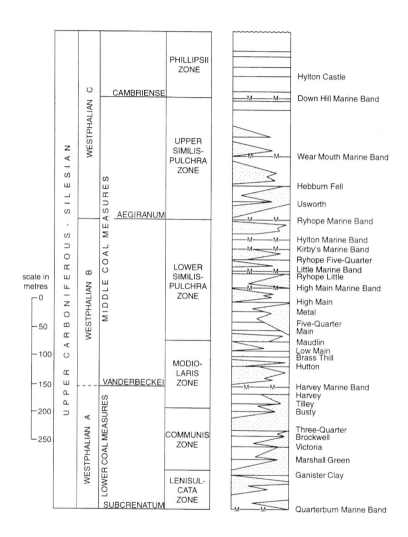

Figure 18. General succession of Westphalian Coal Measures strata in the lower Tyne valley.

Geology of Hadrian's Wall

NEWCASTLE. Little can be seen of the Wall in the urban area of Newcastle. The fort, *Pons Aelius,* defended the important Roman bridge over the the Tyne which was positioned on the site of the present day Swing Bridge. The fort has been excavated under the Keep of the Norman 'new castle' which stands centrally on the site of *Pons Aelius,* but there is little to see. The foundations of the Wall below the pavement are recorded by a plaque on the wall of Neville Hall, the Mining Institute building, at the foot of Westgate Road and a milecastle has been found below the Westgate Road Arts Centre in Black Swan courtyard 600 m to the west. A reconstructed fragment of the milecastle is preserved in the courtyard beside the potters workshop. Westgate Road is the beginning of the 18th Century Military Road from Newcastle to Carlisle built to improve east-west mobility after the Jacobite Rebellion of 1745. The road was built on top of the Wall from milecastle 4 at Newcastle as far west as milecastle 33, so there are few visible remains. At Benwell, the Wall fort of *Condercum* lies astride the Wall, but is covered by buildings and a reservoir. Consolidated remains are outside the fort and consist of a small temple of *Antenociticus* and the Vallum with a crossing point leading to the fort (NZ 216647). The High Main Post, a sandstone between the High Main Coal and the High Main Marine Band (Figure 18), forms a strong feature at Benwell Hill (NZ 210651). The sandstone is exposed in the bank on the south side of the road and the underlying mudstone and coal have also been seen here, hut the section is overgrown. Bruce (1867, p. 119) reports Roman sandstone quarries on both sides of the Wall on Benwell Hill.

At Denton Burn (NZ 202654), 5 km west of Newcastle centre, three short sections of the original Broad Wall and turret 7B are preserved on the south side of the A69 West Road. Two sections of the foundation of the Wall are at Charlie Brown's garage and the larger third section, with the walls of turret 73 standing lm high, lies a little way further to the west (Figure l9). These are the first fragments of Hadrian's Wall visible to the west of *Segedunum* at Wallsend and are well-worth visiting. From here westwards the ditch before the Wall and the Vallum can be made out beside the B6528 road through Throckley to Heddon-on-the-Wall. A 90 m stretch of the Broad Wall foundations is preserved on the south side of the road at Heddon between turret 11B and milecastle 12 (Figure 20) and both the ditch and the Vallum can be made out here.

About 1 km WNW of Denton Burn roundabout the major Ninety Fathom Fault (downthrow N) crosses the line of the Wall (Figure 13). Steeply dipping strata adjacent to the fault were reported in the burn section in the last century, but nothing can be seen now in the urban area. The fault is well-exposed in coastal cliffs on the south side of Cullercoats Bay (NZ 366711). The Coal Measures, particularly sandstones, can be seen in strong scarp features and in the deep denes draining south into the Tyne in the vicinity of Wallbottle, Throckley and Heddon, but there are no continuous sections.

Geology of Hadrian's Wall

Figure 19. *Hadrian's Broad Wall with turret 7B at Denton Burn, Newcastle upon Tyne.*

Figure 20. *Broad Wall over a metre high with a bend in the Wall at Heddon-on-the-Wall. The ring of stones in the foreground is thought to be a post-Roman pottery-kiln.*

Geology of Hadrian's Wall

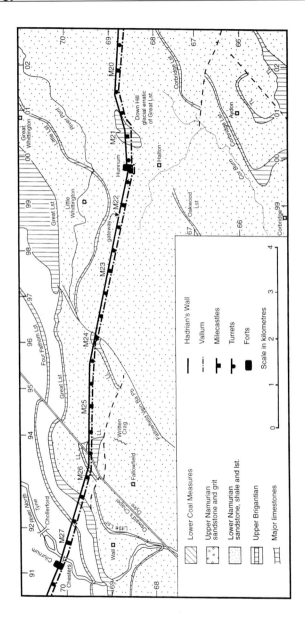

Geology of Hadrian's Wall

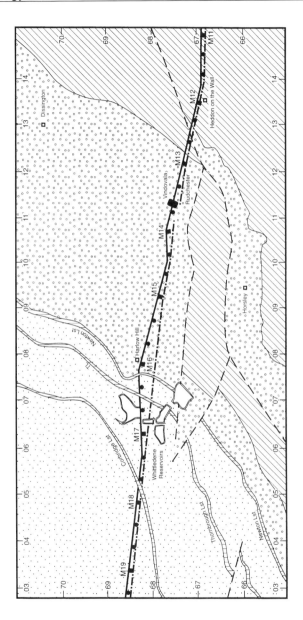

Figure 21. Geological map showing Hadrian's Wall crossing the Namurian outcrop in the eastern sector.

Geology of Hadrian's Wall

HEDDON-ON-THE-WALL. At Heddon the Military Road (B6318) branches WNW from the B6528, the old west road, and follows the line of Hadrian's Wall. The road is built on top of the Wall for most of the 32 km to milecastle 33 with the ditch in the north side of the road and the Vallum on the south. The fort on the Wall at Rudchester (*Vindovala;* NZ 113675) is grass-covered and there is little to see from the road; the land is private with no access. At Harlow Hill, 3.5 km to the west, the Wall, ditch and Vallum are visible near milecastle 16.

From Wallsend to Harlow Hill, the Wall passes over undulating drift covered ground slowly rising towards the west. Westphalian Coal Measures underlie the drift as far west as Heddon-on-the-Wall where Namurian sandstones and mudstones enter in a region cut by strong faults (Figures 21 & 22). Natural outcrops continue to be few, but the thicker sandstone bands in the Coal Measures form features and are near the surface on some ridges, particularly in the Heddon and Horsley area. The Namurian sandstones form more pronounced features with better exposures. Coal and ashes produced by burning coal have been found in the Wall forts and local legend ascribes some ancient, shallow coal workings to Roman mining, but this is unproved (Bruce, 1867, p. 118). Much of the coal extracted by the Romans probably came from natural outcrops of coal seams. At Harlow Hill a Namurian sandstone is well exposed in the disused North Quarry, 500 m north of the Military Road (NZ 078688). Some 250 m to the NE of the sandstone quarry, a disused limestone quarry exposes the Grindstone (= Newton) Limestone (E_2 Namurian). It is one of the highest limestone bands in the Carboniferous to carry a full marine fauna of compound corals and brachiopods. The stratigraphy in this region was described by Hedley and Waite (1929) and a revised sequence and map to the south of the Wall is given on the Geological Survey 1:50,000 Newcastle Sheet (20).

HALTONCHESTERS AND PORTGATE. West of Harlow Hill, the Wall continues to lie below the Military Road, but the ditch and Vallum are both conspicuous and often continuous features. Probable Roman sandstone quarries have been reported at Carr Hill (Bruce, 1867, p. 132). The Military Road leaves the the line of the Wall for a short distance at Down Hill (NZ 006685) and the ditch and Vallum are well preserved near the site of milecastle 21. This hill has been proved to be out of place stratigraphically, being composed of a large raft of Namurian sediments, including the Great Limestone, transported some 3 km SE during the Quaternary glaciation. Extensive quarries on the south side of Down Hill are believed to date from Roman times. Further west, the Wall fort of *Hunnum* at Haltonchesters was built across the Wall, the road passes through the east and west gates. The south side of the fort was extended to the west so that it has an unusual "L" shaped plan. Water supply reached the fort by a buried, stone-built aqueduct from a spring at the head of the River Pont 1.4 km to the NW.

Geology of Hadrian's Wall

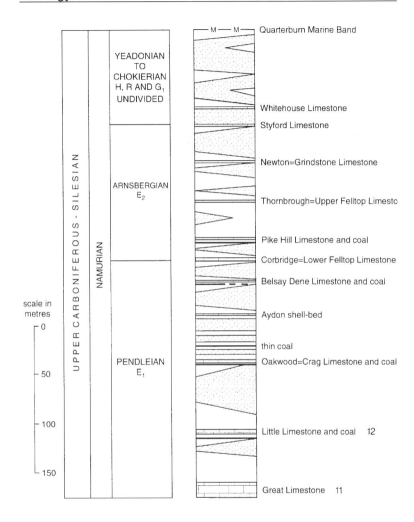

Figure 22. General succession of Namurian strata on the north side of the valley of the River Tyne. Numbered limestones refer to Figure 26.

Geology of Hadrian's Wall

A new roundabout on the Military Road 1 km west of Halton Chesters takes the road away from the line of the Wall and trenches here have proved the Wall below the old road. Near the roundabout, the Roman road of Dere Street crosses the Wall through a gateway called Portgate that controlled passage of travellers passing to the north and south.

CORBRIDGE AND *CORSTOPITUM*. To the south, Dere Street crosses the Tyne at *Corstopitum*, on the west side of the modern town of Corbridge. A fort was established here by Agricola in about A.D. 80 where the Stanegate and Dere Street meet and to guard the bridge over the River Tyne. The fort and surrounding civil settlement were in almost continuous occupation until early in the 5th Century. Water supply to the fort was by an aqueduct running down the hill from the north that terminated in a large cistern with a fountain head of a carved lion with its prey; this is the Corbridge Lion one of the most celebrated Roman carved-stone monuments in Britain. *Corstopitum* was an important supply base and communications centre. It has been extensively excavated and there is much for the visitor to see.

Near Corbridge, the Namurian Thornbrough (= Upper Felltop) Limestone and Corbridge (= Lower Felltop) Limestone are poorly exposed in old workings. More recently, a new exposure of the Corbridge Limestone has been made in a cutting on the north side of the A69 road at Gallowhill (NY 995655). The Corbridge Coal, 0.45 m thick, has been worked in the region and thick seatearth beds (fireclay), particularly below the Corbridge Limestone, supported extensive potteries for the manufacture of glazed pipes and troughs for many years.

FALLOWFIELD FELL. The Military Road lies on top of the Wall west of Portgate for 5.5 km to milecastle 26. Low Namurian sandstones above the Little Limestone are exposed on Fallowfield Fell south of the Wall and were quarried by the Romans. The inscription *"PETRA FLAVI CARANTINI"* - the rock of Falvius Carantinus, was found in a sandstone quarry at Written Crag (NY 937687). The inscription has now been moved for safe keeping to Chesters Museum at Chollerford (Figure 23).

The NE extremity of the Fallowfield Vein carrying galena (lead ore) and witherite (barium carbonate) crosses the wall near to milecastle 24 and was worked from New Engine Shaft 0.8 km NNW of Acomb Cross (NY 931672). This vein is at the eastern side of the Haydon Bridge orefield, a belt of NE-trending mineral veins carrying galena and barium minerals (Dunham, 1990). Several mines produced lead ore during the 19th Century and Settlingstones Mine (NY 845682), which closed in 1968, once produced half the world's supply of witherite. It is interesting that all the mines and productive mineral veins lie just to the south of Hadrian's Wall in Roman territory, but this may be

Geology of Hadrian's Wall

Figure 23. Rock inscription at Written Crag, Fallowfield Fell (Burce, 1867).

Figure 24. The junction of Narrow Wall to Broad Wall at Planetrees, Brunton Bank. Broad Wall foundations, with a conduit below the Wall, had already been laid before the Narrow Wall was built.

Geology of Hadrian's Wall

Figure 25. Sandstone above the Great Limestone in Black Pastures Quarry, Brunton Bank.

simply coincidence. Galena from the Fallowfield Vein contains 4 oz silver per ton of lead. Silver was in much demand by the Romans for their coinage, but there is no evidence of Roman extraction of lead ore hereabouts.

BRUNTON BANK. Near the site of turret 25B, a cross set up on the north side of the Military Road commemorates the Battle of Heaven Fields A.D. 634. St. Oswald, Christian King of Northumbria, defeated Cadwalla, pagan King of Mercia, and brought salvation to northern England and its conversion to Christianity. At Planetrees, just west of milecastle 26, a junction between Broad and Narrow Walls and a conduit below the Wall are seen on the south side of the Military Road (Figure 24). There is an access path, but parking here is difficult. A steep hill (Brunton Bank) leads down the east side of the North Tyne valley and the low Namurian (Pendleian) succession is well exposed in a series of large quarries on the north side of the road. The Great Limestone and shales above, at the base of the Namurian, are exposed in Brunton Quarry (NY 929700) and Cocklaw Quarry (NY 937704). The limestone is some 15 m thick and contains biostromes of *Chaetetes* and algae (Johnson, 1958). The shales above the limestone give way to ripple-marked flags and massive sandstone well seen in Black Pastures Quarry (NY 932699). The sandstone is over 12 m thick in the quarry which is believed to have been initially worked by the Romans (Figure 25); Bruce thought that the stone for the Roman bridge over the River North

Geology of Hadrian's Wall

Tyne came from here. These quarries are all on private land and in particular there is no access to Brunton Quarry (limestone, disused). Black Pastures Quarry is a Northumberland Wildlife Trust Nature Reserve with open access by the track north beside Black Pastures Cottage, half way up Brunton Bank. The working sandstone quarry on the east side is outside the Nature Reserve boundary.

At the foot of Brunton Bank the road heads for the late 18th Century Chollerford Bridge over the River North Tyne. Nearby Chollerton Church has an arcade of circular columns in the nave that are of Roman origin and believed to have come from the fort of *Cilurnum* just to the south.

The Wall and ditch at turret 26A can be seen south of the crossroads at the bottom of Brunton Bank (NY 922698). Access is by a ladder stile at a small parking layby on the A6079, 0.5 km south of the junction with the B6318. A path leads uphill through parkland-pasture to a fine stretch of Narrow Wall on Broad Wall footings running to Brunton turret which is one of the best preserved on the line of the Wall. At the turret the Broad Wall is 2.5 m high and tapers into the Narrow Wall on the west. On the north side of the Wall the ditch is strongly incised.

Geology of Hadrian's Wall

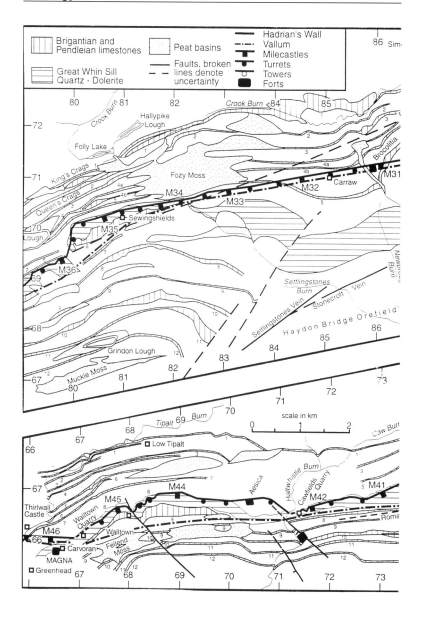

Geology of Hadrian's Wall

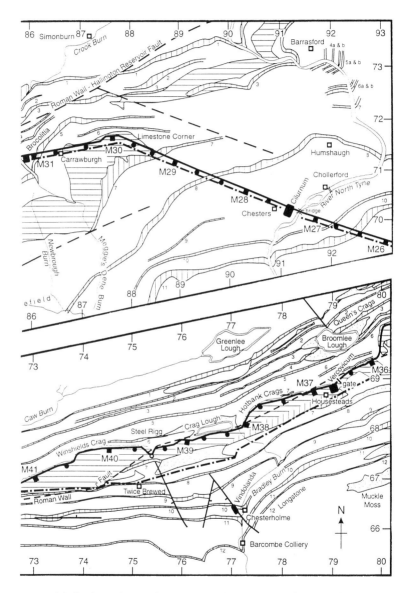

Figure 26. Geological map of Hadrian's Wall in the central sector. The numbered limestone bands 1-10 are Brigantian and 11-12 are Namurian. See stratigraphical successions on Figures 22 and 27.

Geology of Hadrian's Wall

THE CENTRAL SECTOR

The junction between the eastern and central sectors of Hadrian's Wall, chosen at the River North Tyne, is almost coincident with the stratigraphical boundary between the Lower Carboniferous (Dinantian) and the Upper Carboniferous (Silesian) (Figure 26). The stratigraphical boundary lies at the base of the Great Limestone on the evidence of zonal and index fossils found in the north of England (Johnson, Hodge & Fairbairn, 1962; Dunham, 1990). The Dinantian differs from the overlying Silesian in the greater development of limestone in conspicuous cycles of sedimentation (Figure 27). The Yoredale type cycles have the general upward sequence - limestone - mudstone - flags - sandstone - seatearth and coal. They were first described by Phillips (1832) in Wensleydale and are widespread in northern England at the top of the Dinantian.

Ten major sedimentary cycles in the Brigantian Stage can be traced along the line of Hadrian's Wall for 28 km from the River North Tyne to Greenhead. A regional trend causing the limestone bands to split north and east towards the Carboniferous shoreline affects the sequence between the Tynebottom and Scar Limestones in the central sector of the Wall (Figure 27). This has produced uncertainty in lateral correlation, but after suggested revisions (Holliday *et al.* 1975; Frost & Holliday, 1980, figure 15) new findings support the original standard correlation which is used here (Johnson, 1959, 1995; Dunham, 1990).

West of Greenhead, the Asbian Stage Lower Limestone Group and Birdoswald Limestone Group, are composed of less regular depositional cycles in which the clastic component increases down the succession and the thickness of limestone is reduced. The course of the Wall continues over Dinantian strata to the eastern margin of the Carlisle Basin where onlap unconformity of red Permo-Triassic Sherwood sandstone buries the Carboniferous succession along a N-S line near Banks and Lanercost (milecastle 54). Building materials for the Wall change significantly at the line of unconformity; there is no limestone for mortar and no quartzitic sandstone for building-stone to the west. The entrance of red sandstone at the unconformity is taken as the western boundary of the central sector of Hadrian's Wall.

The celebrated quartz-dolerite Great Whin Sill is intruded into the Dinantian succession and forms striking north-facing cliffs across a large part of the central sector of the Wall, from Sewingshields to Greenhead. The Romans found a useful natural barrier and for 16.5 km built the Wall on top of the scarp face. The sill was identified as an intrusion and not a extrusive lava flow in this region of Northumberland by George Tate (1867) who recognised the baked sediments at the top and base of the sill. It is a series of individual sills, varying from 30 m to 67 m in thickness, linked by transgressive steps and injected at successively

Geology of Hadrian's Wall

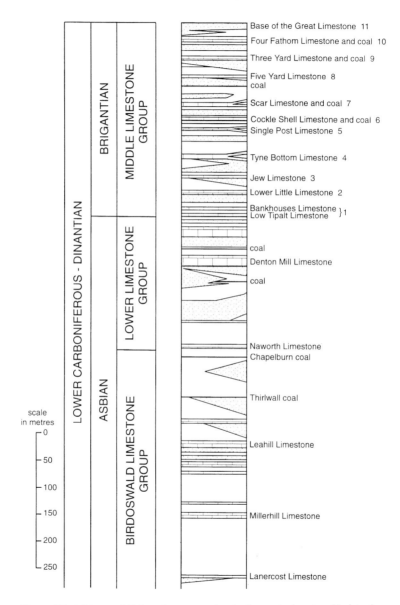

Figure 27. Asbian and Brigantian succession in the central sector, Hadrian's Wall. Numbered limestones refer to outcrops on the map Figure 26.

Geology of Hadrian's Wall

higher horizons when traced from east to west. Thus, near the River North Tyne, the sill crops out north of the Wall in the Jew (= Oxford) Limestone. At Carrawburgh it changes horizon to the Scar and Cockle Shell Limestones and continues at this horizon to Winshields. West of Winshields, the sill lies below the Five Yard Limestone and persists at this horizon to Greenhead where the Whin Sill outcrop diverges from the line of the Wall.

Metamorphism effects the sediments for up to 30 m above and below the sill. Sandstones become quartzites, mudstones develop spotting and turn into porcellaneous whetstones while limestones locally become saccharoidal. Impure limestones produce new metamorphic minerals, such as epidote and garnet. The age of the Whin Sill has been shown to be Silesian and near the top of Carboniferous at 295 Ma. Recent accounts of the Whin Sill and associated dykes in SW Northumberland have been produced by Frost and Holliday (1980) and Randall (1995).

CHOLLERFORD. The Romans carried Hadrian's Wall over the River North Tyne by a bridge with a second bridge, a little way down stream, carrying the Military Way across the river. The abutment of the bridge taking the Wall over the river has been excavated and can be seen on both sides of the river south of Chollerford Bridge (NY 920705). Access to the east abutment is from the B6381 road at Chollerford Bridge by a footpath that follows the left bank of the river to the site; there is space for parking on the verge on the west side of Chollerford Bridge. Near the Roman Bridge, Narrow Wall on Broad Wall foundations leads to a square tower on the bridge abutment. The first Hadrianic Bridge had eight piers, but this was reduced to three piers in later reconstruction when it was enlarged to take both the Wall and the Military Way. The massive east abutment is composed of large dressed sandstone blocks, with lewis holes for lifting, joined by iron cramps set in lead to withstand the force of the river.

On the right bank of the river, the Wall fort of Chesters *(Cilurnum)* is on a terrace above the west bridge abutment and has been extensively excavated. Access is by the B6318 road 0.8 km SW of Chollerford Bridge where a drive on the left leads to a car park at the fort. The fort is built astride the Wall and walls and gateways can be made out beneath turf cover. Some buildings have been exposed and consolidated including the headquarters, the commandants house and the bath house outside the walls of the fort. Water supply came from the North Tyne about 1 km upstream by an aqueduct to a settling tank in the north guardroom of the west gate of the fort. From here the water was led to the houses, the barracks and the bath house. At Chesters Fort there is a celebrated museum with an extensive collection of Roman artefacts and inscribed stones; there are full tourist facilities. Set in pleasant parkland, it is one of the most rewarding forts to visit on the line of the Wall.

Geology of Hadrian's Wall

Figure 28. Quarry in the basal Namurian Great Limestone at Crindledykes, near Chesterholme, Long Stone in the background.

The Dinantian and Lower Namurian succession is fairly well exposed in the River North Tyne (Johnson, 1959; Frost & Holliday, 1980). The basal Namurian Great Limestone (also well exposed near Chesterholme (Figure 28)), crops out in the North Tyne just south of Walwick Grange (NY 907693) and the underlying Dinantian Four Fathom Limestone is found 0.3 km up stream. Both limestone and the intervening Prudhamstone Sandstone, a notable building stone, are exposed in quarries above Fourstones, 2.2 km to the west (NY 885688). Sandstone below the Four Fathom Limestone is well exposed on both sides of the river below the Roman bridge abutments. Upstream the river runs in drift and the Three Yard and Five Yard Limestones are not seen.

Between Haughton Mains (NY 923719) and Barrasford (NY 919731) low river banks and reefs (Figures 27 & 29) expose the succession between the Scar and Tyne Bottom Limestones which is complicated by split limestone bands and intruded quartz-dolerite sills (Johnson, 1959). The section has been described by Frost and Holliday (1980) who use alternative limestone names and correlation, and by Frost (1984) who gives both the standard and alternative nomenclature and reverts to the standard correlation.

The Jew (= Oxford) Limestone crosses the North Tyne SW of Barrasford (NY 915730) and contains reddened algal nodules *(Osagia)* which characterise

Geology of Hadrian's Wall

Figure 29. Exposures in the River North Tyne of the section between the Scar and Tyne Bottom limestones north of Chollerford.

the limestone over much of Northumberland. Below the limestone in the North Tyne the Great Whin Sill is exposed and also at the same horizon in Barrasford Quarry (NY 912743) 1 km to the north. The sill is 30 m thick in the quarry and transgressive through the Oxford Limestone (Randall, 1957; Frost & Holliday, 1980). Some 0.5 km further upstream there are exposures of the Bankhouses and Low Tipalt Limestones at the base of the Brigantian Stage (NY 901734).

Leaving Chesters, Hadrian's Wall lies below the B6313 road up the hill to Walwick Hall. Here the road deviates south on to the edge of the Vallum and both ditch and Vallum can be seen at several places. The stone from the Wall has been robbed hereabouts and the stones can be seen in the old cottage of Tower Tye (NY 891710) and other buildings. Further to the west at Black Carts, a fine stretch of the Wall 0.8 km long with turret 29A and milecastle 30 can be seen on Limestone Bank (Figure 12). The Wall stands up to 1.8 m high and has been excavated and consolidated. The ditch is well preserved before the Wall and the Vallum is equally clear just to the south of the B6318 road. Further south, short rich turf of limestone grassland overlies the Scar Limestone and old quarries and a limekiln can be seen at NY 877709. Sandstone below the limestone has been quarried long ago on Warwick Fell (NY 884708) and it is suggested that the Romans quarried stone here for the Wall (Bruce, 1867) There are extensive sandstone quarries below the Five Yard Limestone at Carr Edge 1 km to the south which may have had a similar origin.

Geology of Hadrian's Wall

LIMESTONE CORNER AND CARRAWBURGH. The Whin Sill is intruded below the Scar Limestone on Limestone Bank and the dolerite-to-limestone boundary is easily traced by changes in the vegetation. It caps the top of the hill at Limestone Corner, the top of Limestone Bank; this is the first place travelling east that the Whin Sill is exposed at the surface. The Wall, ditch and Vallum can be seen and also the Roman Military Way, a road behind the Wall linking the milecastles. It is clearly seen near milecastle 30 where it forms a cambered low ridge below short turf. The original road surface is metalled with light coloured whetstone, baked mudstone from near the Whin Sill contact. The Whin Sill ends abruptly on the east side of Limestone Corner and reappears to the north at Sharply (NY 871724) lower in the succession at the Jew (= Oxford) Limestone (Figure 30). Geophysical investigations have proved that the sill is transgressive and that the two leaves of the sill are not connected by a continuous dyke (Summers *et al.* 1982, p. 114).

Figure 30. View north from Limestone Corner, Carrawburgh. The Whin Sill steps north from the ridge in the foreground to the ridge at Sharpley Farm in the middle distance.

The Romans seem to have made no special provision for cutting the ditch and the Vallum through the hard quartz-dolerite Great Whin Sill. They continued the line of ditch and Vallum from Limestone Bank to Limestone Corner without break attempting to cut each through solid dolerite rock. Large blocks of dolerite were loosened, split and hauled out onto the edge of ditch and Vallum

Geology of Hadrian's Wall

(Figures 31 & 32). But in the case of the ditch some massive blocks, with wedge holes already cut for splitting, were never moved and still lie on the bottom where they were left by the Romans (Figure 33). It is a great tribute to the skill of the Roman engineers that they managed to excavate so much dolerite without the aid of steel and explosives. They only once more, at Walltown, cut the Vallum through Whin Sill dolerite again on the line of the Wall.

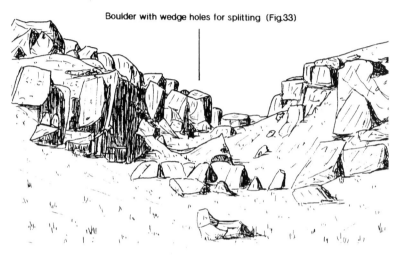

Figure 31. The ditch before the Wall at Limestone Corner cut through the Whin Sill.

Figure 32. The Vallum cut through the Whin Sill at Limestone Corner.

Geology of Hadrian's Wall

Between Limestone Corner and Carrawburgh the Wall is under the B6318 road, but the ditch and Vallum with their mounds of spoil are plain on either side of the road. The Wall fort of Carrawburgh *(Brocolitia)* near milecastle 31 lies behind the Wall and is partly built on the Vallum which had to be filled to make a level space. The buildings of the fort are not exposed and it is grass covered, but the general plan can be made out. There is a civil settlement south and west of the fort including a shrine to the water goddess Coventina and a mithraic temple. Coventina's Well contained many coins when it was excavated during the last century. The site is near the NW corner of the fort, but is now waterlogged and covered with rushes.

Figure 33. Wedge holes for splitting, cut along veinlets in Whin dolerite boulders by Roman workers, in the ditch at Limestone Corner.

The Mithraeum, temple to the sun god Mithras, lies to the SW of the fort and dates from the 3rd Century. It was discovered and excavated in 1949-50 and proved to be a stone building with interior wicker screens and benches which were preserved under a layer of peat (Figure 34). Replicas of altars, statues, wooden roof support, wickerwork, etc. have been installed in the consolidated building and it is a popular tourist attraction; there is a small car park by the fort off the B6318 road. The Mithreaum was abandoned early in the 4th Century and shortly after was inundated by water-borne silt and sand followed by silty peat which covered the building to a depth of 1.8 m. It lies in a small valley where a dam is believed to have formed which allowed the sediment to collect

Geology of Hadrian's Wall

(Richmond & Gillam, 1951). Geologically it seems likely that there was a bog-burst from the peat basin on the north side of the Wall. If the conduit under the Wall was blocked, wet peat would rapidly accumulate against the Wall until the blockage gave way. Peat and water would then flow down the Meggies Dene Burn valley on the south side of the Wall and bury the Mithraeum site (Figure 26).

The succession between the Single Post and Bankhouses Limestones is exposed to the north of Carrawburgh. It is private land and care is necessary to cross the ground without damaging old stone walls and fencing. The Single Post Limestone is exposed in small quarries on either side of the B6318 road 0.2 km east of the fort. The underlying Tyne Bottom Limestone is in two leaves both of which outcrop on the slope north of Tepper Moor Farm. Beds below the Tyne Bottom are reached by the drove road that runs north from the B6318, 0.25 km west of the fort. About 0.7 km down this track a field gate on the left allows access to a quarry and adjacent stone limekiln hidden in a small valley in the centre of the field (NY 857719). This is the Jew (= Oxford) Limestone distinguished by the many red algal nodules *(Osagia)* it contains. These nodules are a persistent feature in Northumberland, but become less conspicuous in Cumbria where the acme of *Osagia* occurs in the Low Tipalt Limestone (Figure 27). Continuing north, a cliff section above Crook Burn exposes the Lower Little (= Greengate Well) Limestone and part of the clastic sequence down to the Bankhouses Limestone, including two coal seams (NY 858723). The Bankhouses Limestone is well exposed in Crook Burn, 1.5 km to the west (Johnson, 1959).

Returning to the road; for 3.6 km west of *Brocolitia* to near turret 33B the Wall lies below the road surface. Carraw Farm, once the rural retreat of the Priors of Hexham, lies across the Wall and the ditch and Vallum are visible on either side of the road with much broken sandstone in the spoil mounds. The works are close together west of Carraw, keeping to the high ground at the south side of Fosy Moss, a large peat basin that had its origin in a glacially-scoured lake. South of the road, ancient sandstone quarries above the Scar Limestone high in Settlingstones Burn (NY 837697) may well have been opened by the Romans. Looking west from milecastle 33, a fine series of escarpments and dip slopes can be seen like waves chasing themselves towards the shore (Figure 35). Most of the Brigantian Stage crops out across these scarps, from the sandstone below the Bankhouses Limestone in the north to the sandstone below the Three Yard Limestone in the south. The central high ground is formed by the scarp and dip slope of the Great Whin Sill. In this region the direction of Quaternary ice movement changed from NW-SE to W-E. The NW-SE moving ice sheet smoothed the geological features, while the W-E moving ice sheet, flowing parallel to the bedrock strike of dipping strata, accentuated the topography with sharp ridges and hollows over hard and soft beds.

Geology of Hadrian's Wall

Figure 34. The Mithraum at Carrawburgh during excavation in 1950.

Figure 35. Bold escarpments north and south of Sewingshields Crags, view west from near to milecastle 33.

Geology of Hadrian's Wall

SEWINGSHIELDS. Owing to the more rugged topography westwards, the Wall, Vallum and Military Road diverge between turrets 33A and B. The Wall continues in the same direction, slightly south of west, following the Whin Sill scarp on to Sewingshields Crags. The Vallum deviates slightly south and keeps to low ground below the dip slope of the Whin Sill, skirting small peat basins. The Roman surveyors seen to have learnt to keep clear of both the Whin Sill dolerite and peaty hollows. The B6318 road heads SW on the well-drained ground of the sandstone above the Scar Limestone. The Five Yard Limestone forms a strong scarp on the south side of the road east of Moss Kennels and is well seen in a small quarry above the road (NY 803695). On the north side of the road, an old quarry (NY 810698), in sandstone below the Five Yard Limestone, was probably opened up by the Romans during the building of the Wall. The sandstone is characterised by crenulate cross-bedding and stone with similar bedding structures occurs regularly in the Wall hereabouts (Frost & Holliday, 1980, p.86).

A ladder stile beside the road (NY 822709), is the start of the footpath along the line of the Wall to Sewingshields Crags. Nearby, turret 33B (NY 821705) has Broad Wall wings and was built before the construction of the narrow curtain Wall. The site of milecastle 34 is covered by a small plantation. The ditch in front of the Wall continues to the milecastle, but beyond this the high Whin Sill scarp made the ditch unnecessary. The Wall is again visible at turret 34A and continues towards Sewingshields Farm. Large rafts of sandstone and limestone are present in the Whin Sill below Sewingshields (NY 810703); and can be seen beside the track leading north from the farm and near to the foot of the escarpment. Further down the track, traces of buildings and fish ponds can be made out below turf in the pasture in front of the Whin Scarp, these are all that remain of the medieval Sewingshields Castle (NY 812705). Folk-lore links this castle and Sewingshields Crags with the legend of King Arthur and his knights who are believed to sleep in a cavern within the crags. The local tale is that the cavern was found by a shepherd lad looking for fox holes below the cliffs. He entered the cave, found the sleeping throng and cut the knot to enter the hall, but failed to blow the horn to wake them up. The sleepers stirred and the shepherd fled in terror; no cave is known.

The footpath by the Wall passes north of Sewingshields Farm through a plantation of fir trees and continues west on the crest of the Whin scarp. The Wall has been excavated and consolidated on Sewingshields Crags and can be followed past a Saxon burial cist, built against the Wall, to milecastle 35. Much Whin Sill dolerite is incorporated in the core of the Wall and the milecastle together with some baked Cockleshell Limestone which overlies the dip-slope of the sill. The Wall continues with breaks to turret 35A and the summit O.S. column (315 m above O.D.). It is a good view-point for the southern margin of

Geology of Hadrian's Wall

the Kielder Forest. The Wall can also be seen continuing westwards on the top of the scarp of the Great Whin Sill towards Housesteads Fort. Broomlee Lough is conspicuous below the crags and to the north a fine series of escarpments are visible between the Asbian King's Crags, below the Low Tipalt Limestone, and the Brigantian Single Post Limestone (Figures 36 & 37). The bold sandstone scarps of King's Crags, Queen's Crags and Dove Crag may well have been quarried by the Romans and one of the old quarries on Queen's Crags has reported wedge holes for splitting the stone. Another Roman sandstone quarry lies above Greenlee Lough (NY 775697) in the King's Crag Sandstone and there is a Roman temporary camp nearby. A Roman limestone quarry and limekiln has also been found.

Continuing west along the course of the Wall, a break in the Whin Sill scarp at Busy Gap is inferred to be where the Hallington Reservoir - Roman Wall Fault displaces the Whin Sill outcrop (Figure 26). This elongate strike-slip fault has been proved in part by geophysical investigation (Frost, 1984). The sill changes horizon at Busy Gap from below the Cockle Shell Limestone to the Scar Limestone as it rises slightly in the succession westwards.
The Wall was protected by a ditch across the flat bottom of Busy Gap. This wide break in the Whin Sill scarp was notorious in times gone by for cattle-rustlers and hence its name. Over King's Hill and Kennel Crags, the Wall continues on

Figure 36. Massive cross-bedded sandstone above the Lower Little Limestone at Queen's Crags showing glacial plucking and the development of coarse scree.

Geology of Hadrian's Wall

Figure 37. *North-facing scarps of limestone, sandstone and Whin dolerite south and east of Broomlee Lough. Both sandstone and limestone were quarried by the Romans in this region.*

the Whin Sill scarp, but has been reduced to a field boundary. Approaching Knag Burn, the Wall resumes its full width and continues to Knag Burn Gate, one of the gates through the Wall for trading with the northern tribes. Two guard houses on either side of the gate and double doors indicate that traders could be checked in both directions, going north or south.

HOUSESTEADS FORT *(VERCOVICIUM).* Built on the site of turret 36B on the west side of Knag Burn, high on the scarp of the Whin Sill and overlooking magnificent scenery (see upper figure on the back cover), Housesteads is the most popular fort on Hadrian's Wall (Figure 38). A large car park and full tourist facilities has been developed beside the B6318 road below the fort (NY 794684) and from here, a rough track leads mainly up hill to the fort, civil settlement and museum. The perimeter walls, gateways and much of the interior of the fort have been uncovered and consolidated and there is much to see. Water supply by aqueduct was difficult at this high and exposed site, so as much water as possible was collected from roofs of buildings in large stone tanks. Otherwise water had to be carried to the fort from Knag Burn where there is a well near the bath house. At a late stage, the guard-room at the east gate was turned into a coal store where a large quantity of local coal was found. There are workable coal seams below the Three Yard, Four Fathom and Little Limestones in the vicinity

Geology of Hadrian's Wall

of Housesteads. The Four Fathom Coal, 0.45 m thick, exposed less than 1 km to the south, is the nearest to the fort. The Little Limestone Coal, a better seam and up to 0.6 m thick, outcrops 1.6 km to the south and, at the east end of Grindon Lough, it was possibly worked by the Romans.

South of Housesteads the Vallum is cut through the Scar Limestones taking a course well behind the Wall keeping clear of the Whin Sill and peat basins. Further west it diverges to the edge of the B6318 road, 600 m south of the Wall, to miss a wide peat basin at the foot of the dip-slope of the Whin Sill. At Housesteads the Whin Sill transgresses through the Scar Limestone on the upper contact with baked shale (whetstone) can be seen in Knag Burn to the east of the fort (NY 792689). A Roman limekiln has been excavated above Knag Burn.

Figure 38. Housesteads Fort (Vercovicium) *sited on the crest of the Whin Sill escarpment, view of the dip-slope from the south.*

The view south from Housesteads is over the drainage basin of Knag Burn including Grindon Lough. The southern margin of the basin is a strong sandstone scarp above the Namurian Little Limestone along which runs the Stanegate Roman Road. The basin has internal drainage into active potholes in the Four Fathom Limestone NW of Grindon Lough (NY 797679). The water reappears, according to local report, in Bradley Burn near Chesterholme. The Great Limestone (Figure 28) is exposed in quarries at Crindledykes where it was burnt for lime up to the 1950's.

Geology of Hadrian's Wall

Figure 39. Vindolanda *Fort, Chesterholme, seen from the east. Drift covered ground south of the Wall with bedrock exposures in the valley bottoms.*

There are strong sandstone scarps north and south of Housesteads, the best building stone coming from the north and inferior stone from the south of the Wall according to Hodgson (1822). The most celebrated locality is near Long Stone on Thorngrafton Common (Figure 28) where a Roman bronze arm purse containing 63 coins was found hidden in the spoil of an ancient sandstone quarry (NY 782665); the treasure, called the Thorngrafton Find, is now in the Chesters Museum.

CHESTERHOLME *(VINDOLANDA)*. From Long Stone looking west the Stanegate Road can be seen passing Chesterholme and the fort of *Vindolanda*, one of the original Stanegate forts (NY 770663). The plan of the fort can be made out and part of the civilian settlement has been excavated (Figure 39). Full sized stone and turf replicas of Hadrian's Wall have been constructed at *Vindolanda* for the many visitors to the site. Access is by the Stanegate Road from either the east or the west. It is well sign-posted and has full tourist facilities with car park, shop and a good museum. Beside the Stanegate Road, near to where it crosses Bradley Burn, a Roman milestone stands in its original situation (Figure 40).

Geology of Hadrian's Wall

Figure 40. A Roman milestone standing in its original position by the Stanegate Road at Chesterholme.

Brackies Burn, NW of *Vindolanda*, exposes the Three Yard Limestone and overlying shales (NY 763666). Water supply to the fort came from this burn and was led by aqueduct to stone water tanks. Large ironstone nodules (iron ore) from the shales above the Three Yard Limestone were worked in quarries NW of *Vindolanda* during the early 19th Century (NY 767666). No Roman working of the ironstone is known, though at Corbridge, iron smelted from the Redesdale ironstone of the North Tyne valley has been reported. Bradley Burn above Chesterholme exposes part of the succession between the Three Yard and Four Fathom Limestones.

CRAG LOUGH. West of Housesteads, the Wall continues on the crest of the Whin Sill scarp for 4 km to Steel Rigg over National Trust land with paths, but rough going. Milecastle 37 west of Housesteads is well preserved and shows Broad Wall tapering to Narrow Wall on either side. This much visited stretch of the Wall continues over Cuddy's Crag and Hotbank Crags, passing turrets 37A and B, to Hotbank Farm. There are fine views from Cuddy's Crag eastwards along the Wall to Sewingshields (back cover) and from above Hotbank Farm westwards over Crag Lough to Peel Crags and Winshields Crags (Figure 41). The ditch before the Wall resumes at Hotbank and can be traced past the site of milecastle 38 to the break in the Whin Sill scarp called Milking Gap. Crag Lough, a celebrated glacial lake in an ice-scoured basin, is on the west side of

Geology of Hadrian's Wall

Figure 41. North facing scarp of the Whin Sill at Crag Lough. The Wall (foreground) continues westwards following the line of crags.

Figure 42. Castle Nick and Sycamore Gap in the Whin Sill escarpment. Dip-slope of the sill ends in the peat basin, Vallum in foreground.

Geology of Hadrian's Wall

Figure 43. Peel Gap Tower mainly built in Whin dolerite. A later addition to the Wall that protects a wide gap in the Whin Sill scarp.

the gap at the foot of the Whin Sill cliffs called Highshield Crags. The Wall continues to follow the Whin Sill scarp with some well preserved stretches. Two narrow gaps in the scarp break the line of precipitous cliffs. Sycamore Gap contains a mature tree and the other, called Castle Nick, contains milecastle 39 which has been excavated and consolidated (Figure 42). Peel Crags follow and are high dark cliffs of roughly columnar quartz-dolerite with a glacial spillway from Crag Lough at the foot. Good stretches of Wall continue from Peel Crags over the spillway to Steel Rigg. The low ground of the glacial spillway through the Whin Sill escarpment was defended by a late tower on the Wall called the Peel Gap tower. It is noteworthy for the almost exclusive use of Whin Sill dolerite in facing stones and core of the walls (Figure 43).

STEEL RIGG. A small car park with some tourist facilities has been set up at Steel Rigg (NY 751677) by the National Park, 1 km north of the B6318 at Once Brewed. East of the car park the Wall is well preserved and a walk of 100 m gives a fine view of the Whin Sill escarpment and the glacial spillway at the foot of the cliffs. Crag Lough, the glacial lake, is in the middle distance with Hotbank Crags in the background. At Once Brewed, the Northumberland National Park Visitor Centre (NY 751668) has full tourist facilities; a popular Youth Hostel and the Twice Brewed Inn are nearby.

Geology of Hadrian's Wall

East and West Twice Brewed Farms and the Twice Brewed Inn form a notable district. According to the old farmer of East Twice Brewed, met in the field almost 50 years ago, the name comes from an English King on campaign with his army against the Scots who stopped at an inn beside the footpath SE of the farm. He asked for ale and was given a weak brew that was unacceptable, so he commanded the innkeeper to brew it again. Probably this tale relates to Edward I who travelled in the district with his army and sheltered at Bradley Hall, a nearby farm (NY 778675), on the 6th and 7th September 1306 as recorded in the Chronicle of Lanercost quoted by Bruce (1867, p.207). King Edward visited various settlements in the region before spending the winter of 1305-7 at Lanercost Priory (NY 556637).

A track leads west from Steel Rigg along the line of the Wall with the ditch continuing to milecastle 40 where the high cliffs of the Whin Sill scarp resume. Winshields Crags (NY 740675) is the highest point on the Wall between Tyne and Solway (375 m above O.D.). Here, the Wall is well preserved in two stretches between milecastles 40 and 41. Glacial meltwater has cut deep channels into the Whin Sill on Winshields Crags up to a height of $c.320$ m at the top of Green Slack (Figure 44), where a flat bottomed spillway runs SE down the dip slope (NY 742675). Further west, Lodhams Slack (NY 738672), at $c.300$ m, runs in the same direction and forms a break in the Whin Sill scarp. A minor transgession in the Whin Sill takes place in the vicinity of Lodhams Slack where

Figure 44. Green Slack glacial spillway cut through Whin Sill quartz-dolerite high on Winshields Crags.

Geology of Hadrian's Wall

it is suggested that the western end of the Hallington Reservoir - Roman Wall Fault crosses the Whin Sill escarpment (Frost, 1984). Other small transgressions of the sill occur at Thorny Doors (NY 722668) and Cawfields Quarry (NY 715666) to the west.

CAWFIELDS. A fine stretch of the Wall runs for almost 1 km from Shield-on-the-Wall Farm west to Cawfields milecastle 42. The Vallum, returning to its normal position behind the Wall after skirting the Whin Sill and peat basins, is well developed at the foot of the dip slope of the Whin Sill east of Cawfields Quarry. The Vallum ditch with spoil mounds, two to the south and one to the north, are clearly seen (Figure 45). South of the Vallum on the side of Haltwhistle Burn there are Roman temporary camps and a Stanegate Fort that was abandoned when the forts were moved to the line of the Wall. Old quarries for limestone, ironstone and coal are visible near the outcrop of the Three Yard Limestone (NY 718633) and old shafts after coal are located by Haltwhistle Burn. An inscription by Legion VI was found in a sandstone quarry above the Three Yard Limestone on the west side of the fort (NY 715662), but was destroyed by further 19th Century quarrying. Near the Milecastle Inn (NY 716660) there are quarries in the Four Fathom Limestone (Figure 46) on the north side of

Figure 45. The Vallum running in a straight line east from Cawfields Quarry at the foot of the dip slope of the Great Whin Sill.

Geology of Hadrian's Wall

Figure 46. Quarry in the top Brigantian Four Fathom Limestone beside the B6318 road near Cawfields.

B5318 road and the Great Limestone on the south of the road. There is a car park and picnic area at Cawfields (NY 713666) with basic tourist facilities provided by Northumberland County Council. Cawfields Quarry gives exposures of the Great Whin Sill in two leaves with intervening baked shale and sandstone. The upper sill extends only a short distance east of the quarry and is thin, while the lower sill is thicker and continuous eastwards. To the west of the quarry the sills are covered by some 20 m of glacial drift in the buried valley of Haltwhistle Burn. Geophysical investigations indicate that the sill is continuous westwards with only a small transgression to the east of Cawfields Quarry (Summers *et al.* 1982). Milecastle 42, just east of the quarry (Figure 8), is built on the dip slope of the Whin Sill and is well preserved with walls 2.4 m thick and gateways formed of massive masonry (NY 716667).

HALTWHISTLE BURN. The gorge of Haltwhistle Burn, south of the B6318 road, gives the best section through the Lower Namurian (Pendleian) rocks of the region (Figure 47). The burn and the River South Tyne are believed to have been part of the River Irthing drainage in pre-glacial times and they flowed westwards through Greenhead and Gilsland. The present easterly Tyne drainage appears to have developed during the retreat stage of the last glaciation (Trotter & Hollingworth, 1932, p. 167). The Haltwhistle Burn section starts at the Great Limestone and provides an almost continuous succession up to the sandstone above the Oakwood (= Crag) Limestone. The footpath down the burn starts at the white

Geology of Hadrian's Wall

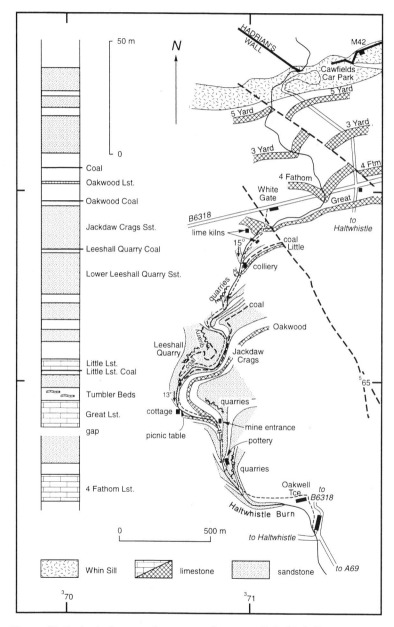

Figure 47. Geological map and sequence of strata at Haltwhistle Burn.

gate just west of the cottage on the south side of the B6318 road (NY 714559). The path is well-marked and is the old narrow gauge railway track from Cawfields Quarry to Haltwhistle. From the gate the field slopes steeply down to the burn where there is a footbridge and an exposure of the Great Limestone. The limestone is about 7 m thick and is overlain by interbedded limestone and calcareous mudstone bands called the Tumbler Beds. Down stream, past old limekilns, the sandstone above the Great Limestone is exposed on the east bank with fine detail of sedimentary structures picked out by wind erosion. Past the sandstone cliff, a chimney and engine block mark the position of a colliery that worked the Little Limestone Coal. The Little Limestone that overlies the coal is exposed in the cliff behind the colliery. It is about 5 m thick of shaly limestone and calcareous shale and is fossiliferous containing particularly brachiopods and bryozoa. A footbridge further down the path leads to the west bank of the burn and to quarries in massive sandstone interbedded with mudstone.

In Leeshall Quarry, on the west bank (Figure 47), there is a good exposure of two sandstone bands separated by a coal seam 0.25 m thick resting on 2 m of seatearth. Below Leeshill Quarry the path returns to the east bank and the top of the upper Leeshall Quarry Sandstone can be seen in the stream at the picnic site. Downstream, the Oakwood (= Crag) Limestone crops out just above water level in the west bank and the exposure can be traced for 20 m downstream to just above the wooden footbridge. The limestone is impure with a high clay content and is usually only about 2 m thick, but it is a significant marker horizon in this part of the succession which is dominated by sandstone and mudstone. The shale and sandstone above the Oakwood Limestone are exposed in a large quarry set back from the stream on the east bank and below which there are ruins of another colliery, This drift mine worked a thin coal above the limestone, but in its later days produced fireclay for the pottery a short distance down stream. The pottery closed a few years ago, but the main product, glazed pipes, are still to be seen. The path continues on the east bank of the stream, beside old sandstone quarries, to Haltwhistle.

GREAT CHESTERS *(AESICA)*. A footpath leads west from Cawfields to Great Chesters Farm and the Wall fort of *Aesica.* Deep drift in the buried valley of Haltwhistle Burn gives way westwards to the dip slope of the Great Whin Sill, but the fort and Great Chesters Farm lie on the crest of a drift ridge (Figure 26). *Aesica* was built entirely behind the Wall and was sited to defend the low ground of the Haltwhistle Burn valley. Though the fort has been excavated at various times the masonry has not been consolidated and it is mainly under turf, but the plan with the walls and gateways can be made out. The west wall is best exposed; the guard rooms at each side of the gate and two corner turrets can be seen clearly. Within the fort the outlines of barrack blocks and part of the vault of the strong room in the headquarters building can be seen; an altar remains

Geology of Hadrian's Wall

standing in the SE corner of the fort. Milecastle 43 was dismantled when the fort was built. Water supply to the fort came from Caw Burn, a northward continuation of Haltwhistle Burn, by a winding aqueduct some 9.5 km long that is marked on the O.S. maps. Between turrets 43A and B, on the west side of a small plantation beside Cockmount Hill, there is a Roman milestone that has been recycled to form an unusual field gatepost.

A long southerly dip slope of the Whin Sill and overlying Five Yard Limestone develops west of Great Chesters and continues to Walltown (Figure 48). The Vallum, close to the Wall at Cawfields, is diverted south to miss the Whin Sill and peat basins and lies on higher ground well away from the dip slope of the sill. West of *Aesica,* fine runs of the Wall pass milecastles 44 and 45.

WALLTOWN. English Heritage has consolidated over 300 m of the Wall and turret 45A on Walltown Crags (NY674664). It is one the best preserved sections of Wall in its entire length with sufficient height remaining to give some idea of what it must have been like during the Roman occupation (see front cover). Turrets 44B and 45A are well preserved and at the latter the turret clearly preceded the Wall with Broad Wall wings on either side quickly changing to Narrow Wall. Access is signposted from the B6318 road, but there is only limited parking space. Some of the gaps in the Whin Sill escarpment near Walltown, called the Nine Nicks of Thirlwall, are caused by faults (NY 680666). The Whin Sill is here in the Five Yard Limestone which can be seen in small quarries above the car park at Walltown. A coal seam below the Five Yard Limestone up to 1.8 m thick has been mined on the north side of the Wall and to the south the Little Limestone Coal is 1.5 m thick at Fellend (NY 682655). It has been extensively worked in the area and is still being mined. Crags of massive sandstone on both sides of the Wall may have been quarried by the Romans and one quarry at Fellend lies within a Roman Marching Camp. The Vallum, skirting peat basins south of Walltown, crosses the narrow outcrop of the Whin Sill following the course of a small stream in a drift filled valley, just to the west of the termination of the Whin escarpment (Figure 26).

Walltown Quarry, at the end of the Whin Sill escarpment, destroyed turret 45B and more than 300 m of Wall (Figure 49). The quarry closed in 1978 and since then the site has been landscaped by Northumberland County Council using spoil from the new Greenhead bypass. The sill and underlying sediments are exposed in the quarry and a popular Hard Rock Trail leads visitors round the outcrops. A visitor centre has been built here and there are plans to reconstruct the Wall and turret 45B on the quarry floor. Beyond the quarry the Wall has been robbed of stone, but it can be seen continuing westwards.

The Whin Sill runs SW under drift in the Tipalt Burn valley and the glacially

Geology of Hadrian's Wall

Figure 48. Fellend Moss looking east to Cawfields and Winshields Crags.

Figure 49. Hadrian's Wall at turret 45A, Walltown (foreground), destroyed eastwards by past quarrying in the Great Whin Sill.

Geology of Hadrian's Wall

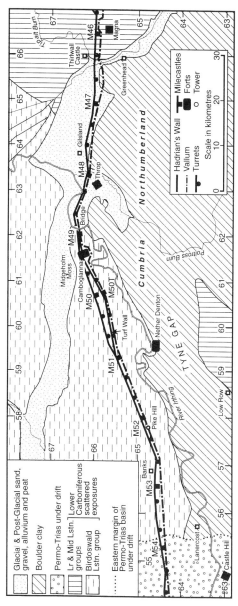

Figure 50. Geological map showing the course of Hadrian's Wall over Asbian and Brigantian strata between Greenhead and Banks.

Geology of Hadrian's Wall

eroded bedrock escarpments of the east are lost below thick glacial deposits of the Tyne Gap (Figure 50). A boring near Greenhead proved 44.2 m of boulder clay. Rounded drumlin topography is associated with the Tyne Gap, a major pass through which ice travelled eastwards during the Quaternary glaciation. Lake District erratics form a boulder train from west to east with Shap Granite boulders found in milecastle No.48 at Gilsland, above Twice Brewed (Johnson, 1952; NY 743682) and by Crook Burn below Teppermoor (NY 858721).

CARVORAN. The west end of the Whin Sill escarpment and the east side of the Tipalt valley were defended by the fort of *Carvoran (Magnis, Magna)*, now owned by the Vindolanda Trust. It is a Stanegate Fort sited at a strategic cross roads of Stanegate with the Maiden Way from Kirkby Thore and Alston to the south. The fort lies behind the Wall and they are separated by a small peat basin. The Vallum lies close to the Wall north of the fort, keeping clear of the peaty hollow, and then diverges south on each side to regain its normal position. *Magna* has not been excavated and little can be seen except the NW corner because the ground has been ploughed. The site has, however, yielded many inscriptions and artefacts including quantities of cinders from burnt coal. It will produce much of interest when modern excavation takes place. Carvoran Farm (NY666658) contains the Roman Army Museum which displays artefacts and reconstructions of the Roman garrisons on Hadrian's Wall; there is a car park and full tourist facilities.

The Wall and milecastle 46 can be seen north of *Carvoran,* but west of this the Wall is lost, though the ditch can be seen as far as turret 46A. The stone facing blocks from the Wall and no doubt stone from the core, have been used in farm buildings. In addition, the nearby medieval Thirlwall Castle, now in ruins to the north of the Wall, is built of Roman stones. The ditch before the Wall and the Vallum run parallel down the slope to Tipalt Burn. On the west side of the valley, the Wall can be seen in two short stretches on either side of the B6318 road (NY 657661; Figure 7) and on the west side of the road the Wall, ditch and Vallum are conspicuous in a long stretch past milecastles 47 and 48.

Between *Carvoran* and Gilsland, the course of the Wall crosses the Asbian and Brigantian Lower and Middle Limestone Group successions of well-developed Yoredale cyclothems. Much of the area lies on the west side of the Blenkinsop Boundary Fault in strata that dips up to 50° SE and the Lower and Middle Limestone Groups are condensed into a belt 2.3 km wide, but the beds are poorly exposed. The best section through the sequence is in Tipalt Burn, Greenhead, which exposes much of the Middle Limestone Group and the upper part of the Lower Limestone Group (Trotter & Hollingworth, 1932; Johnson, 1959).

Poltross Burn, Gilsland, also gives good sections between the Bankhouses and Cockle Shell Limestones above Temon (NY 617639) and near the Naworth

Geology of Hadrian's Wall

Limestone at Throp (NY 630655). The burn drains north to the River Irthing which in turn flows west into the River Eden (Figures 50 & 54). It is the start of westerly drainage into the Irish Sea and just west of the primary watershed between the Tyne, easterly drainage and the Eden, westerly drainage. The line of the watershed is irregular just to the east of Gilsland where tributaries draining east and west overlap in the region of the Tyne Gap. It is noteworthy that adjacent to the watershed, Poltross Burn is the county boundary between Northumberland and Cumbria (Figure 50).

GILSLAND. According to early accounts the Wall crossed the incised valley of Poltross Burn with an arch, but nothing remains of this. Milecastle 48, on the west side of the burn, has been excavated and consolidated and shows the exterior and interior plan well. Access to the milecastle is by a footpath starting from near Gilsland railway station. Poltross Burn is here thickly wooded, but gives discontinuous bedrock exposures near the horizon of the Naworth Limestone. Above the burn, Throp is an early Stangate fort built of turf and timber that was reoccupied in the 4th Century (NY 631660).

Gilsland was a watering place and spa town during the 18th and 19th centuries where visitors came to take the sulphurous and chalybeate spring waters that were believed to have medicinal properties. The springs are in the south end of the Gilsland Gorge, 1.5 km north of Gilsland (NY 634675). The sulphurous spring gives off carbon dioxide and hydrogen sulphide and a little further north the chalybeate spring water contains ferrous sulphate. In both springs the iron and sulphur are thought to be derived from the adjacent pyritous shales (Trotter & Hollingworth, 1932, p.41-42).

At Gilsland the line of the Wall passes on to the Asbian Birdoswald Limestone Group, some 426 m of limestone, mudstone, sandstone and coal seams in broad coarsening-upwards cycles (Figures 27 & 50). The succession is well exposed in the River Irthing which flows south of west and joins the River Eden near Carlisle. Beds near top of the Birdoswald Limestone Group are exposed for 9 km in the gorge of the Irthing with cliffs up to 18 m high in some places. Detail of the succession between Birdoswald Fort and Gunshall (NY 586649) is given by Trotter and Hollingworth (1932, p.34, fig.3, section A). The lower part of the Birdoswald Limestone Group is arenaceous with thick sandstones predominating, but in upward succession limestone enters and the measures are more evenly balanced with limestone, mudstone and sandstone. Twelve coal seams have been recognised, but only the Thirlwall, up to 1 m thick, and the Chapleburn, some 0.6 m thick, have been worked. The River Irthing section north of Gilsland has fine exposures of the Asbian Birdoswald Limestone Group which have been described in detail by Day (1970, p.125-131). Spreads of glacial sand and gravel to the east and west of Gilsland are kames and deposits of a glacial lake formed during the early retreat stages of the of the last glaciation (Trotter, 1929, p.587).

Figure 51. Turret 48A at Gilsland with Broad Wall wings.

The Wall can be seen with only short breaks for about 1 km on the south and west of Gilsland past turrets 48A and 48B. It is consolidated Narrow Wall on Broad Wall foundations and it can be traced with gaps to turret 48A which has Broad Wall wings (Figure 51). Just west of the turret, the River Irthing has cut southwards at a large bend and both the ditch and the Wall have slipped away down the steep bank. Past the river bend, the ditch, Wall and Vallum can be followed to turret 48B at Willowford Farm where visitors pay the farmer to proceed further.

WILLOWFORD. The fine stretch of Wall continues past Willowford Farm across an alluvial terrace to the site of the Roman bridge over the River Irthing, called Willowford Bridge (NY 622665). The situation is much like Chesters Bridge at Chollerford, the Wall had to cross the Irthing and initially a narrow bridge carrying only the Wall was constructed. Later it was widened and a tower was built to guard the crossing. A complicated abutment in massive masonry was erected to protect against river erosion and was rebuilt three times, once after flood damage and twice later during alterations. Only the eastern abutment can be seen because the river has cut into its west bank and carried away the west end of the bridge. Two water piers of the bridge are known and are buried in alluvium. River erosion of the boulder clay cliff has produced a steep slope

Geology of Hadrian's Wall

above the west bank of the Irthing and here the Wall has disappeared. There is no connection between Willowford Bridge and the west bank of the Irthing leading to Birdoswald Fort and to continue west along the line of the Wall you have to retrace steps to the Gilsland Road, take the B6318 over Irthing Bridge and then the first turning on the left to Birdoswald; the route is well signposted.

At Willowford Bridge the Broad Wall ends and only Narrow Wall is found to the west. Here, the early Hadrianic Wall was constructed of turf and later replaced by Narrow Wall in stone. On the top of the high west bank of the River Irthing, milecastle 49 (Harrow Scar) was originally built of turf and later reconstructed in stone. The Narrow Wall (stone) is well preserved and deviates slightly north to join the north wall of Birdoswald Fort (see back cover). Behind it the earlier Turf Wall produces a low ridge across the fields and takes a straight line from turret 48B to the west gate of Birdoswald.

BIRDOSWALD *(BANNA, CAMBOGLANNA).* The site of Birdoswald Fort (NY 615663) is on high ground over-looking the River Irthing on the south and protected on the north by Midgeholme Moss and the deep valley of Midgeholme Beck. Here, the first Wall was of turf, probably because it was quick to build. Slightly later, possibly without undue haste, this was replaced by stone which may well have come from the Irthing gorge. A Roman sandstone quarry at Coombe Crag (NY 591650), west of Birdoswald and just south of the Wall, contains inscriptions of the names of soldiers who worked the stone during the consulship of Faustinus and Rufus (A.D. 210). At about this time the Turf Wall was replaced by stone under Severus (Table 1); much of the stone used in the wall between Birdoswald and Banks is similar to the Coombe Crag stone. The Irthing gorge, with its accessible reserves of sandstone and limestone, was wisely retained south of the Wall.

Birdoswald Farm, made of Roman stones, lies within the perimeter of the fort and is a visitor centre. There is a car park and full tourist facilities. Entrance is from the north of the fort and behind the farm the line of the walls and gateways can be made out, but much of the interior is still covered by turf. Recent excavations have exposed the granaries near the west gate of the fort and remains of a Saxon hall building with flagged floors. Birdoswald Fort was built on the site of turret 49A with the original ditch excavated in front of it. The turret was taken down and ditch was filled to produce a level site for the fort at a slightly later date. First a cavalry fort, with three gates in the north wall, it was sited to guard Willowford Bridge over the Irthing and the road *(Fanum Cocidi)* leading north to Bewcastle Fort. Birdoswald was converted to an infantry fort, with only one gate in the north wall, when the Turf Wall was abandoned and the Stone Wall was constructed.

Geology of Hadrian's Wall

The masonry at Birdoswald Fort is some of the best on the Wall. Fine grained buff sandstone, similar to the Coombe Crag stone, allowed a fine ashlar finish and tight joints of adjacent facing blocks. The ramparts and gateways show a considerable thickness of apparently undisturbed stonework and are the best surviving examples on the Wall (Figure 52).

Figure 52. South gate and rampart of Birdoswald Fort showing well dressed massive masonry, apparently undisturbed.

A minor road leads from Birdoswald to Lanercost running parallel and just south of the Narrow Wall (stone). The ditch before the Wall can be made out together with a short length of the Wall near to turret 49B. Slightly further south, the Turf Wall diverges and then converges on the Narrow Wall which it joins at milecastle 51 (Wall Bowers). The Turf Wall can be seen crossing the farm track south of Appletree (NY 597656) where the ditch amd Vallum are also clearly visible. Originally the Turf Wall continued westward as can be seen at turret 51A (Pipers Sike, NY 588653) where the original turret was retained when the Narrow Wall was constructed and they were not bonded together (Figure 53). Lea Hill turret 51B (NY 584651) shows the same features and has masonry standing 8 courses high. Banks East turret 52A (NY 575647) is another original Turf Wall turret with adjoining Narrow Wall (Figure 9). It is maintained by English Heritage and has a small car park and information panels.

A footpath leads back beside the road to Pike Hill (NY 577648) where the remains of a Roman signal tower have been excavated and consolidated on the

Geology of Hadrian's Wall

Figure 53. Pipers Sike turret 51A, an original Turf Wall turret.

south side of the road. The tower predates the Wall and is set obliquely to it so that a double bend in the Wall is formed. Much of the tower extended over the road and has been lost, only a fragment remaining. The Wall can be seen at Hare Hill (NY 564646) just west of Banks standing some 3 m high (Figure 10). It was the highest length of original Wall, but consolidation this century required almost complete rebuilding of the facing stones. Both the ditch and the Vallum are visible across Craggle Hill and west to below Garthside. Geologically the eastern margin of the Carlisle Basin crosses the line of the Wall to the west of Banks. The entrance here of Permo-Triassic red sandstone, concealed by thick glacial deposits on the line of the Wall, is taken as the west end of the central sector of Hadrian's Wall.

Geology of Hadrian's Wall

THE WESTERN SECTOR

The boundary between the central and western sectors of Hadrian's Wall is taken at the eastern margin of the Carlisle Basin (Figure 54). It is marked by a conspicuous change in the colour of the bedrock, from the grey and buff of the Lower Carboniferous Birdoswald Limestone Group to the red and red-brown of the Triassic Sherwood (= St Bees) Sandstone (Figure 55). The colour change of the strata is very striking and effects also the colour of the drift deposits and soils. A more rounded lowland topography is developed over the less resistant red sandstone and mudstone. For the builders of Hadrian's Wall there was no difficulty in excavating the ditch before the Wall and the Vallum behind it, but accessible places where red sandstone could be obtained for building were limited to locations well away from the Wall. The Turf Wall was probably constructed because it was quick to build compared to building a stone Wall which was not completed here until late in the second century (Table 1).

CARLISLE BASIN. During Permo-Triassic times subsidence in the Carlisle and Vale of Eden basins took place in two phases. In the initial phase, the Permian Brockram Conglomerate and Penrith Sandstone deposition was fault controlled and restricted to the centre of the basins. Some 0.75 km thickness of Permian aeolian sands and conglomerate were laid down before a change to regional uniform subsidence took place towards the end of the Permian with the transgression of the Zechstein (= *Bakevellia*) Sea over western Britain and the Irish Sea. The Eden (= St Bees) Shales were laid down with thick beds of gypsum and anhydrite and thin bands of magnesian limestone (Figure 55). There is biostratigraphical evidence to correlate tentatively the Belah Dolomite of the Vale of Eden with the Seaham Beds (EZ3) Upper Permian of the Durham sequence in eastern England (Burgess & Holliday, 1979). The Eden Shales spread beyond the margins of the Permian Carlisle Basin with onlap over the Permian and underlying Carboniferous deposits all round the basins. In succession the Triassic Sherwood Sandstone (= St Bees and Kirklinton Sandstones) further onlapped over the margins of the basins and the Mercia Mudstones (= Stanwix Shales) possibly did likewise. Post-Triassic erosion of the Carlisle Basin has restricted the outcrop of the Kirklinton Sandstone and the Mercia Mudstone to the centre of the basin and the Upper Permian Eden Shales only occur as a narrow outcrop at the basin margin. There are some 700 m of Sherwood Sandstone and 300 m of Mercia Mudstone in the Carlisle Basin. The absence of coarse-grained alluvial facies at the eastern margin of the basin, suggests that both sandstone and mudstone onlapped far to the east and that the source of sediment was also distant to the east.

A complex unconformity, with successive onlap east of the Triassic succession, forms the eastern margin of the Carlisle Basin. It was mistakenly interpreted as a

Geology of Hadrian's Wall

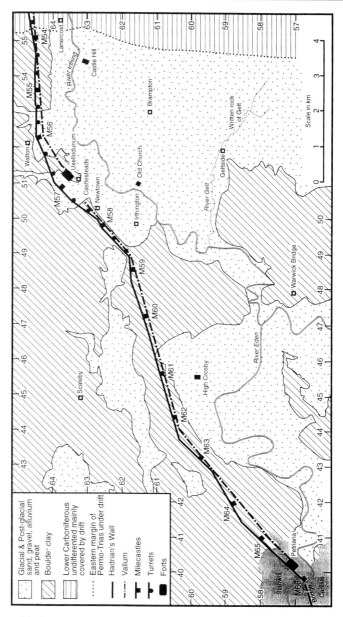

Figure 54. Geological map showing the course of the Wall over drift-covered country between Banks and Carlisle.

Geology of Hadrian's Wall

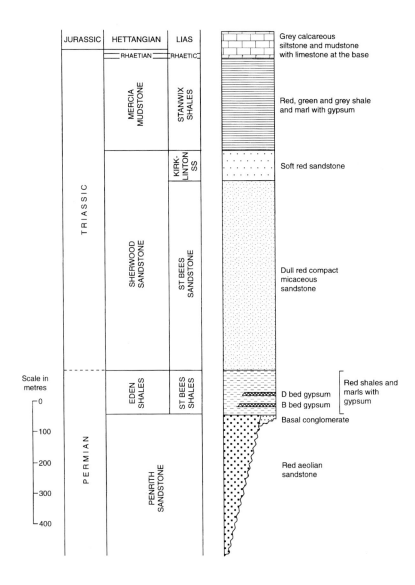

Figure 55. General succession of Permian, Triassic and Jurassic strata of the Carlisle Basin, western, sector of Hadrian's Wall.

Geology of Hadrian's Wall

faulted boundary by some early worker's and the name Red Rock Fault has been used. The Pennine Fault was presumed to extend north near the eastern margin of the Carlisle Basin (Trotter & Hollingworth, 1932), but no evidence of this fault has been found north of the Stublick Line at the NW corner of the Alston Block. The unconformity is concealed below drift over much of the eastern margin of the basin. Beds near to it are exposed at Quarry Beck, south of Lanercost Bridge (NY 548621) where a breccia 0.5 m thick underlies some 21 m of Eden Shales, but the base is not seen (Trotter & Hollingworth, 1932, p. 136). The junction of Trias with Carboniferous is seen in a tributary of Burtholme Beck ENE of Garthside (NY 552648) and near the line of the Wall. Here, Permo-Triassic basal breccia is seen resting on Carboniferous beds. Further north about 10 m of the breccia is exposed in the Cam Beck at Kirkcambeck (NY 535690).

West of Carlisle an outlier of late Triassic, Rhaetian and Lower Jurassic Hettangian Lias occupies the centre of the Carlisle Basin (Figure 64). It is poorly exposed, but detail was provided in 1989 by three boreholes drilled by British Gypsum Ltd (Ivimey-Cook *et al.*, 1995). Red-brown followed by grey mudstone of the Mercia Mudstone were proved below interbedded grey shale and limestone containing Rhaetian fossils (Figure 55). Grey shale and limestone that followed contained *Psiloceras planorbis* and other molluscs indicative of the Lower Jurassic Lias (Hettangian). Some 13 m of fossiliferous Rhaetian beds and 70 m of Lias were proved in borehole SB1 *(ibid.)*.

BANKS TO WALTON. A change in topography from the crisp upland crags of the Central Sector of the Wall to the more rounded lowland hills to the west begins at Greenhead with the thick glacial deposits of the Tyne Gap. This change becomes more marked westwards, particularly west of Banks, where the indurated Carboniferous sequence gives way to the softer red sandstones of the Carlisle Basin. Thick glacial drift masks the underlying bedrock outcrop west of Banks and few exposures of the red-brown Sherwood Sandstone occur, none on the line of the Wall. In this area the course of the Wall was chosen along ridges of boulder clay on the north side of the Irthing valley. They are better drained and give a better foundation than the adjacent lower ground covered with glacial gravel, recent alluvium and peat. Two north-south tributaries of the River Irthing, Burtholme Beck and the King Water, with valley bottom alluvium had to be negotiated.

Westwards on the line of the Wall, the last place where it is seen standing with a solid core and partly clad with facing stones is at Hare Hill, Banks (Figure 10). From here to the west end of the Wall against the Solway Firth only traces of the Wall and its ditches and mounds are to be seen. The missing facing stones of the Wall are conspicuous, only in the sense that they have been recycled into Saxon,

Geology of Hadrian's Wall

medieval and later buildings. For example, at the 12th Century Lanercost Priory 1.5 km SW of Banks (NY 556637), part ruin part splendid parish church, Roman masonry, both red Triassic and buff Carboniferous sandstone have been used almost exclusively. The medieval builders must have plundered dressed stone from several kilometres of the Wall to construct the great priory church and the surrounding cloisteral buildings. Assuming the church was built from east to west, the builders must have taken the well-cemented buff Carboniferous sandstone from the east of Banks first, in preference to the softer Triassic red sandstone which appears in quantity at the west end of the church and higher in the walls. The volume of Roman dressed stone in the remaining priory walls gives no doubt as to why Hadrian's Wall has almost disappeared in this region.

Figure 56. *Ditch with line of the Wall following the hedge on the right,. View SE of Garthside, near Banks, looking east. There are loose stones from the Wall core in the bottom of the hedge.*

On both sides of Burtholme Beck, the Vallum can be seen with the ditch before the Wall on the east side of the stream. Broken red sandstone, the core of the Wall, is visible in the bottom of a hedge running west up the hill from the stream to the SE of Garthside (NY 547644; Figure 56). The course of the Turf Wall ran slightly to the north of the Stone Wall here, but little can be made out on the surface. The Wall is present on the west bank of the King Water at Dovecote Bridge (NY 527644), where a short length of red sandstone wall was visible until 1983. This interesting section was deteriorating so rapidly owing to weathering and disintegration of the stone that it has been covered with clay soil

Geology of Hadrian's Wall

Figure 57. Red sandstone Wall covered by protective clay soil at Dovecote Bridge, Walton.

and now appears as a low grassy ridge (Figure 57). The site is enclosed and is protected by English Heritage. Westwards, above the King Water, Walton is a pleasant green village sited on milecastle 56. The King Water is in drift at Dovecote Bridge, but to the north, upstream near Moorguards (NY 580674), there are good exposures of the Lower Carboniferous Holkerian and basal Asbian sequence, the Craig Hill Sandstone Group of Trotter and Hollingworth (1932, p. 31).

WALTON TO NEWTOWN. The course of the Wall continues over undulating drift covered ground mainly used as pasture. The ditch before the Wall can be seen in places and the core of the Wall can form a low ridge below turf or is present as broken sandstone in hedge bottoms (Figure 58). Broken red sandstone of the Wall core is well seen below a hedge beside the footpath along the line of the Wall east of Newtown (NY 502629; Figure 59) and in the village an undamaged red sandstone facing-stone from the Wall lies on the top of a low dry-stone wall. The Cam Beck valley, south of the Wall is covered with alluvium, but upstream to the north, red Kirkliston Sandstone, high in the Sherwood Sandstone succession, is exposed for 2.4 km. Basal Permo-Triassic breccias occur in the beck at Kirkcambeck and to the NE near Parkgate Bridge (NY 556705) fossiliferous limestone and mudstone of the Lower Carboniferous Cambeck Beds (Arundian) are well exposed (Day, 1970, p. 77).

Geology of Hadrian's Wall

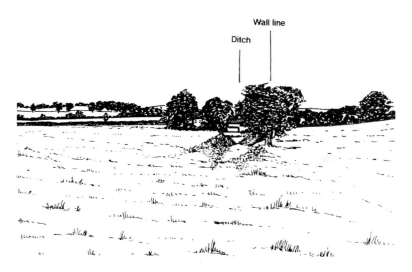

Figure 58. Ditch and Wall line at the hedge (middle distance) ploughed out in the foreground field. View east near turret 57B at Newtown.

The Cam Beck flows through the grounds of Castlesteads House to the south of the Wall and beside a fort that has been named *Uxellodunum*, but this is uncertain. The fort was sited in a strong defensive position on a ridge above the beck with steep ground to the north and is unique in lying south of the Wall and within the Vallum. The fort and the Wall hereabouts were destroyed when the house was built in the 1790's. The fort was excavated in 1934, but is in private grounds with no access. South of the River Irthing the forts of Castle Hill and Old Church are on the east-west Stanegate Road behind the Wall.

The gorge of the River Gelt, a tributary of the River Irthing, lies some 5 km south of the Wall and is an important section through the red Sherwood Sandstone which contains Roman quarries. The gorge is about 24 m deep with vertical sandstone cliffs up to 15 m high, exposing some 200 m of strata. The sandstone is usually massive, but splitting into flagstones with partings of red-brown shale at intervals of about 6 m (Trotter & Hollingworth, 1932, p.139). Celebrated inscriptions (Figure 60) carved on the rock by Roman quarrymen occur at the crag called Written Rock of Gelt (NY 525589) where members of Legion II *Augusta* and Legion XX *Valeria Victrix* worked under Mercatius in about A.D. 207 (Bruce, 1867, p.82). The date fits with the renovation and rebuilding of the Wall under Severus at the beginning of the 3rd Century (Table 1). Another Roman quarry in the gorge at Pigeon Crag, 0.8 km up stream of the Written Rock, has an inscription by men of Legion VI.

Geology of Hadrian's Wall

Figure 59. Remains of the red sandstone core of the Wall in a hedge line at Newtown.

Figure 60. Rock inscription at the Written Rock of Gelt (from Bruce, 1867, p.82).

Geology of Hadrian's Wall

NEWTOWN TO STANWIX. West of Newtown, the topography is more subdued and soon approaches the smooth expanse of Carlisle Airport. The course of the Wall takes the crest of a low boulder clay ridge on the north side of the Irthing valley for 12.5 km to Stanwix, Carlisle. The ditch before the Wall can be seen at several places west of Newtown and particularly between Oldwall, Bleatarn and Wallhead, milecastles 59 to 61. At Oldwall the ditch is visible on both sides of the lane and can be followed east and west on good footpaths (Figure 61). The walls of a barn beside the footpath east contain many Roman red sandstone facing stones. At Wallhead, milecastle 61, the ditch can be seen east of the farm and is almost continuous to Bleatarn. In the bottom of the hedge on the south side of the lane, outside the farm, inconspicuous broken red sandstone seems to be a last vestige of the core of the Wall. Birky Lane runs in a straight line for 1.5 km SW of Wallhead. It lies on the berm with the ditch on the north side of the lane and traces of the Wall in the hedge on the south (Figure 62). The bottom of the hedge consists of sandy loam containing pieces of red sandstone and is the much weathered residual of the Wall core. The lane continues through Walby to Wallfoot where it joins the B6264 road. Turn right towards Carlisle and right again at the nearby round-about and continue for 300 m to a parking place in a lane on the right of the road. Brunstock Park lies on the west side of the road and a fine view of a long straight stretch of the ditch before the Wall can be seen crossing the park-land (Figure 63).

Figure 61. The ditch, berm and the line of the Wall along the hedge on the left. View west on the west side of the lane at Oldwall.

Geology of Hadrian's Wall

Figure 62. The ditch, berm and line of the Wall on the side of Birky Lane. View west of Wallhead looking west. The hedge bottom on the south side of Birky Lane contains red sandstone, the remains of the Wall core.

Figure 63. The ditch before the Wall crossing Brunstock Park, Tarraby, Carlisle.

Geology of Hadrian's Wall

There are no exposures of bedrock near the Wall west of Newtown, but 6.5 km south at Wetheral, a gorge in the River Eden exposes some 150 m of massive Sherwood Sandstone divided by only a few thin shale bands. This is the largest and best exposed section of the Sherwood Sandstone near to the Wall (Trotter & Hollingworth, 1932, p.139). The stone has been quarried and a Roman inscription by men of Legion XX *Valeria Victrix* was recorded by Bruce (1867) low down on the cliff on the west side of the River Eden, near the three caves called St Constantine's Cells (NY 467537; Figure 4). Although this inscription cannot be dated, Legion XX was probably procuring stone at Wetheral at the same time as their work in the Gelt Gorge at the beginning of the 3rd Century.

Carlisle was an important military and civilian centre during the Roman occupation. On the north side of the River Eden at Stanwix, *Petriana* was the largest and most important fort on the Wall. Founded in Hadrian's time, the garrison was a crack regiment of 1000 cavalry, the *Ala Petriana*. who were based here to the end of the 3rd Century. The fort is sited on more elevated ground above the River Eden, but there is little to see because it is completely built over. The early antiquarians visiting the fort in the 18th Century could find little detail and most of the stone was probably recycled into new buildings in medieval times. Excavations have proved the line of the Wall that forms the north wall of the fort and some of the internal buildings. The Wall runs SW from the fort to the confluence of the River Eden with the River Caldew where there was a bridge. The bridge abutments have not been found, but dredging in the Eden has produced a collection of red sandstone blocks that were part of the piers of the Roman bridge; these stones have been gathered together and placed where the bridge abutment is believed to have stood on the west side of the River Eden in Bitts Park (NY 396567).

Luguvalium, the civil settlement with fort and garrison, was on the south side of the River Eden about 365 m south of *Petriana.* It was one of Agricola's original forts on the Stangate Road at the junction with the north-south route from Chester to Scotland and about it developed a thriving commercial centre from about A.D. 71. The site of *Luguvalium* is also completely built over and the Roman masonry has been recycled into new buildings certainly from the Middle Ages onwards. Roman artefacts and inscriptions from this and other Cumbrian sites are well displayed in the new Tullie House Museum in Castle Street, Carlisle.

WEST OF CARLISLE. The line of the Wall follows higher ground on the south bank of the River Eden to the NW of Carlisle to the village of Beaumont and then runs west to the Solway Firth at Burgh-by-Sands and Watch Hill (Figure 64). Here it reaches the flat coastal marshes and continues along the shore, using any available higher ground, to the end of the Wall at Bowness-on-Solway. The Vallum is at varying distances from the Wall since its course is

Geology of Hadrian's Wall

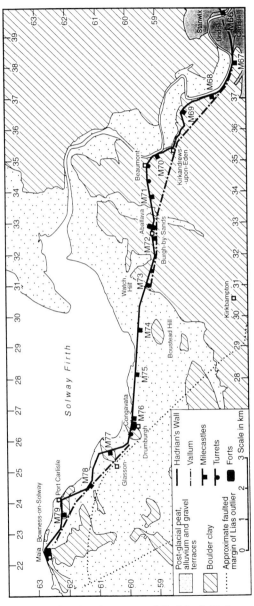

Figure 64. Geological map showing the course of Hadrian's Wall over drift covered country between Carlisle and Bowness-on-Solway.

Geology of Hadrian's Wall

straight between the major southerly bends, while the Wall follows higher ground. South of the Wall the Stangate Road runs to *Luguvalium*, but may continue west, near the line of the present B5307 road, to the Roman fort at Kirkbride (NY 235565).

The region is covered with reddish-brown glacial boulder clay about 4 m thick over the Lias subcrop, but in places between 9 m to 15 m are present with up to 90 m over the crest of some drumlins (Dixon *et al.*, 1926). Along the line of the Wall, boulder clay stands out as low rounded drumlins at Beaumont, Watch Hill, Drumburgh, Glasson and Bowness, surrounded by post-glacial and recent marine alluvium and gravel terraces. The boulder clay is derived from the north and contains Criffel and Dalbeattie granite erratics.

Overlying an eroded surface of boulder clay, a submerged forest peat bed is known at several places round the coast of Cumbria (Dixon *et al.* 1926, p.76). On the line of the Wall it was reported 0.8 km NW of Glasson in excavations for the Carlisle Canal (Bruce, 1857, p.302). A prostrate forest is described with trunks of trees lying in a N-S direction with some trunks 0.6 m to 0.9 m high still standing in the position of growth. The foundation of the Wall is said to pass obliquely over the forest bed at about 1 m above the level of the trees. The peat bed has been reported on the Solway shore near Glasson where a freshwater swamp mud is separated from the underlying boulder clay by a thin layer of stoneless blue clay (Walker, 1966). Pollen analysis of the organic mud gives a position in zones III and IV, 9000 B.C. to 8000 B.C., lower Flandrian, and the overlying marine alluvium must date from about 7500 B.C. The peat bed is exposed on the coast south of Bowness at Cardurnock and near Beckfoot, Silloth (Dixon *et al.* 1926). It is believed to have formed during a post-glacial (Flandrian) low sea-level period which ended with a sea-level rise that caused destruction of the terrestrial forest and onset of deposition of marine alluvium. Similar woody peat deposits on the NE coast of England have also been dated at 9000 B.C. to 8000 B.C. (Smith & Francis, 1967, p.248).

Extensive marine alluvium (warp), pale yellow, brown and grey sandy clay, overlies the forest bed and forms most of the low ground on the south shore of the Solway Firth; it has been divided into lower and upper warp terraces (Dixon *et al.*, 1926). The course of the Wall passes over coastal warp terraces for most of the distance from Watch Hill, Burgh-by-Sands to Bowness-on-Solway.

The marine alluvium terraces were formed at periods of relatively high sea-level during the Flandrian. Complicated sea-level changes at this time were the result of glacio-isostatic adjustment of the crust following removal of ice over the British Isles combined with eustatic change of sea-level from melting of Continental ice sheets. These processes and their combined effect in the

Geology of Hadrian's Wall

area of the Irish Sea has been recently described by Lambeck (1996).

There are no bedrock exposures on the south shore of the Solway Firth and exposures of solid deposits are only found over the Lias plateau to the south where the surface outcrops have been revised recently (Ivimey-Cook et al., 1995). There are no exposures of Sherwood Sandstone in the region and the Romans must have carried stone from quarries at least 8 km to the south. Bruce (1867, p.300) suggests that the Romans went to Howrigg Quarry about 13 km to the south for facing stones for the Wall and to Stonepot Scar, a quarry on the north shore of the Solway Firth, for the stones of the core of the Wall. Unfortunately there is no Wall left in place to test these suggestions. Hadrian's Wall was seen by Bruce (1867, p.303) in his younger days, in the earlier part of the 19th Century, near Port Carlisle standing 2 m high with its rugged and weathered core as hard as rock. He reflects that no advantage has resulted from its destruction, it occupied little space and it served the purpose of a fence. In the absence of rock exposures suitable for quarrying building stone, Hadrian's Wall has been used extensively for this purpose. The complete disappearance of the Wall west of Carlisle is thus explained. The churches of Beaumont, Burgh-by-Sands and Great Orton and the castle of Drumburgh as well as humbler dwellings are constructed of Roman stones.

CARLISLE TO BURGH-BY-SANDS. Hadrian's Wall is built on boulder clay NW of Carlisle and runs on higher ground on the west bank of the River Eden. The Wall has not survived, but the ditch before the wall is visible in places, particularly at Grinsale where it forms a hollow and grassy banks (NY 369579). The Vallum has been proved in a few places in this region. The course of the Wall follows the River Eden to Beaumont which lies on an elongate drumlinoid knoll. St Mary's Church, partly constructed of Roman stones, is on the site of turret 70A at the centre of a pronounced angle in the Wall that now runs west to Burgh-by-Sands. An inscribed stone built into a wall near to the church is a building stone of the 5th cohort of Legion XX and suggests that they were the construction team for the stone curtain Wall hereabouts.

In Burgh-by-Sands, St. Michael's Church, a fortified medieval building almost entirely constructed of Roman stones, is in the centre of the Wall fort of *Aballava* (Figure 65). The fort covers an area of about 5 acres and served to guard the ancient ford across the Solway Firth to Scotland; it is known from inscriptions to have had a strong garrison during the 2nd and 3rd centuries.

Burgh village is built over the fort and there is little to see. The Wall was excavated nearby in 1986 when the Turf Wall was found on a cobble foundation and the later Stone Wall was built on the same line in sandstone (Johnson, 1989, p.36). Below Burgh, on the edge of the Solway marshes, Edward I of England died on

Geology of Hadrian's Wall

Figure 65. St. Michael's Church, Burgh-by-Sands mainly constructed of Roman stones from the surrounding fort of Aballava.

7th July, 1307 waiting for the right tide to cross the ford with his army to campaign in Scotland. A monument was erected on the site in 1685 and rebuilt in 1803; it now suffers subsidence problems.

BURGH-BY-SANDS TO BOWNESS-ON-SOLWAY. The course of the Wall continues west from Burgh over boulder clay to the low drumlinoid knoll of Watch Hill (milecastle 73) overlooking the marshy Solway shore. Descending the hill to the salt marsh, the Wall crosses Burgh and Eastern marshes on 1st terrace marine alluvium and recent alluvium. There is no surface evidence of the course of the Wall and the foundations must be deeply buried in the marsh.

At Drumburgh the Wall rises from Eastern Marsh at milecastle 76 on to a further low drumlinoid knoll on which the Wall fort of *Congavata* was built (Figures 64 & 66). It was a small fort covering only 2 acres and lies on top of the knoll with a wide view on all sides. Originally a turf and timber fort that was bonded into the Turf Wall, it was replaced in stone, at a slightly smaller size, probably at the same time as the Stone Wall was built in about A.D. 160. The stones from the fort of *Congavata* and the Wall seem to have disappeared before the end of medieval times being used to construct Drumburgh Castle, a fortified manor house, in the centre of the village. It has a fine flight of steps leading to the main entrance and a large Roman altar beside the door (Figure 67). The junction

Geology of Hadrian's Wall

Figure 66. The coastal Eastern Marsh on first and second terrace marine alluvium. Low drumlinoid hills of Glasson and Drumburgh in the distance. View west from near Watch Hill, Burgh-by-Sands.

Figure 67. Drumburgh Castle constructed of Roman stones. A large Roman altar stands beside the door.

Geology of Hadrian's Wall

Figure 68. Low coastal cliff of boulder clay at the north end of the low drumlinoid hill on which Glasson is built.

between boulder clay and marine alluvium can be seen on the shore below Drumburgh.

From Drumburgh, the course of the Wall continues westwards over 1st terrace marine alluvium to Glasson, which is situated on another elongate drumlinoid knoll of boulder clay just west of milecastle 77 (Figure 64). Marine erosion of the north face of the drumlin has exposed boulder clay which contains many erratics (Figure 68). The Wall runs back on to 2nd terrace marine alluvium west of Glasson and heads inland to the south of the present road to milecastle 78. It changes direction here to NW and continues over the marine terrace to Port Carlisle at turret 78B. Port Carlisle is on a further low drumlinoid mound of boulder clay and protected from the sea by a small headland. It is the seaward entrance to the Carlisle Canal which allowed seagoing ships to reach Carlisle between 1823 and 1840. Remains of locks are still present at the entrance to the canal basin, but the site is much altered because in the 1850s, a railway was laid on the course of the old canal.

The site of turret 78B, on the north side of Port Carlisle, marks a change in the line of the Wall and it runs inland, on the south side of the road over 2nd terrace marine alluvium, to Bowness-on-Solway (Figure 69). The Vallum comes close to the Wall where it diverges away from the coastal marshes and runs in straight

Geology of Hadrian's Wall

Figure 69. Eroding second terrace of marine alluvium on the Solway shore west of Port Carlisle.

Figure 70. The Solway Firth from the Banks, the north rampart of Maia *Fort, Bowness-on-Solway.*

Geology of Hadrian's Wall

lines cutting off bends in the Wall. It has been proved at several places between Glasson and Bowness.

Bowness-on-Solway is built on a headland formed by a low drumlinoid mound of boulder clay and other drumlins lie to the east, west and south. It is a natural look-out with unrestricted views over the Solway Firth and out to sea. The present village covers the Roman Wall fort of *Maia,* the western terminus of Hadrian's Wall. It is an attractive village with a narrow entrance formed by terraced houses close to the edge of the road and no pavements. Nothing can be seen of the fort below the houses, but a map on the west wall of the inn at the road junction to Kirkbride, shows the relation of *Maia* Fort to the present village.

The fort was originally built of timber, turf and clay and was later rebuilt in stone. Excavations have proved the south and west walls and the north wall, continuous with the line of Hadrian's Wall, runs near to the top of the boulder clay sea cliff overlooking the Solway (Figure 70). It was a large fort covering an area of about 7 acres and it was occupied well into the 4th Century. A large civil settlement developed south of *Maia* along the sides of the road to the fort at Kirkbride.

Maia was built in a position to the west of the last or most seaward ford over the Solway Firth. There are several fords where the Solway can be crossed at low water and the Romans were determined to restrict unauthorised travellers at the Cumbrian coast. Thus the curtain Wall was extended as a barrer to Bowness beyond the westernmost Solway ford. Roman defences against intruders from the sea continued south down the coast of Cumbria for 41 km in a series of milefortlets and towers much like the milecastles and turrets of Hadrian's Wall (Johnson, 1989, p.62). There was no curtain wall, however, and the milefortlets were built of turf and timber, and not replaced by stone. Forts were built at Kirkbride, Beckfoot and Maryport, but the west end of Hadrian's Wall, the northern frontier of the Roman Empire, was at *Maia,* Bowness-on-Solway.

Geology of Hadrian's Wall

FURTHER READING

BREEZE, D.J. & DOBSON, B. 1978. *Hadrian's Wall.* Harmondsworth: Pelican, 324pp.

BRUCE, J. C. 1867. *The Roman Wall: A description of the Mural Barrier of the North of England.* 3rd Edn., London: Longmans, xv+466pp.

BURGESS, I.C. & HOLLIDAY D.W. 1979. Geology of the country around Brough-under-Stainmore. *Mem. Geol. Surv. G.B., Sheet* 31, 131pp.

CROW, J. 1995. *English Heritage Book of Housesteads.* London: Batesford, 126pp.

DAY, J.B.W. 1970. Geology of the Country around Bewcastle. *Mem. Geol. Surv. G.B., Sheet* 12, 357pp.

DIXON, E.E.L., MADEN, J., TROTTER, F.M., HOLLINGWORTH, S.E. & TONKS, L.H. 1926. The Geology of the Carlisle, Longtown and Silloth District. *Mem. Geol. Surv. G.B.* Sheet 17, 113pp.

DUNHAM, K.C. 1990. Geology of the Northern Pennine Orefield, Volume 1 Tyne to Stainmore (2nd Edn). *Econ. Mem. Brit. Geol. Surv.,* 299pp.

FROST, D.V. 1984. New information on the Dinantian stratigraphy and structure of Tynedale, Northumberland. *Proc. Yorkshire Geol. Soc.,* **45**, 45-49.

FROST, D.V. & HOLLIDAY, D.W. 1980. Geology of the country around Bellingham. *Mem. Geol. Surv. G.B.,* Sheet 13, 113pp.

HEDLEY, W.P. & WAITE, S.T. 1929. The sequence of the Upper Limestone Group between Corbridge and Belsay. *Proc. Univ. Durham phil. Soc.,* **8**, 136-152.

HODGSON, J. 1822. Observations on the Roman Station of Housesteads. *Archaeologia Aeliana[1],* **1**, 263-320.

HOLLIDAY, D.W., BURGESS, I.C. & FROST, D.V. 1975. A recorrelation of the Yoredale Limestones (Upper Viséan) of the Alston Block with those of the Northumberland Trough. *Proc. Yorkshire Geol. Soc.,* **40**, 319-334.

HOPKINS, T. 1993. *Walking the Wall.* Newcastle upon Tyne: Keepdate, 148pp.

IVIMEY-COOK, H.C., WARRINGTON, G., WORLEY, N.E., HOLLOWAY, S. & YOUNG, B. 1995. Rocks of the Late Triassic and Early Jurassic age in the Carlisle Basin, Cumbria (north-west England). *Proc. Yorkshire Geol. Soc.,* **50**, 305-316.

JOHNSON, G.A.L. 1952. A glacial erratic boulder of Shap Granite in south Northumberland. *Geol. Mag.,* **89**, 361-364.

JOHNSON, G.A.L. 1958. Biostromes in the Namurian Great Limestone of Northern England. *Palaeontology,* **1**, 147-157.

JOHNSON, G.A.L. 1959. The Carboniferous stratigraphy of the Roman Wall district in western Northumberland. *Proc. Yorkshire Geol. Soc.,* **32**, 83-130.

JOHNSON, G.A.L. (Ed.) 1995. Robson's Geology of North East England. *Trans. Nat. Hist. Soc. Northumbria.,* **56**, 225-391

Geology of Hadrian's Wall

JOHNSON, G.A.L., HODGE, B.L. & FAIRBAIRN, R. 1962. The base of the Namurian and of the Millstone Grit in north-eastern England. *Proc. Yorkshire Geol. Soc.*, **33**, 341-362.

JOHNSON, S. 1989. *English Heritage Book of Hadrian's Wall.* London: Batsford, 143pp.

JONES, J.M. 1957. The geology of the coast section from Tynemouth to Seaton Sluice. *Trans. Nat. Hist. Soc. Northumb.,* **16**, 153-192.

LAMBECK, K. 1996. Glaciation and sea-level change for Ireland and the Irish Sea since the end of Devensian/Midlandian time. *Jl geol Soc. Lond.,* **153**, 853-872.

LAND, D.H. 1974. Geology of the Tynemouth district. *Mem. Geol. Surv. G.B.,* Sheet 15, 176pp.

PENNINGTON, W. 1978. Quaternary Geology. In MOSELEY, F. (Ed.) *The Geology of the Lake District.* Yorks. Geol. Soc. , Occas. Publ. No.3, 284pp.

PHILLIPS, J. 1836. *Illustrations of the geology of Yorkshire Part 2. The Mountain Limestone District.* London: John Murray, 253pp.

RANDALL, B.A.O. 1959. An intrusive phenomenon of the Whin Sill, east of the North Tyne. *Geol. Mag.,* **96**. 385-392.

RANDALL, B.A.O. 1995. Carboniferous and Tertiary igneous rocks. In JOHNSON, G.A.L. (Ed.) Robson's Geology of North East England. *Trans. Nat. Hist. Soc. Northumb.,* **56**, 317-329.

RICHARDS, M. 1993. *Hadrian's Wall, Vol. 1 The Wall Walk.* Milnthorpe: Cicerone, 206pp.

RICHMOND, I.A. & GILLAM, J.P. 1951. The Temple of Mithras at Carrawburgh. *Archaeologia Aeliana⁴*, **29**, 1-92.

SMITH, D.B. 1994. Geology of the country around Sunderland. *Mem. Geol. Surv. G.B.,* Sheet 21, 161pp.

SMITH, D.B. 1995. Permian and Triassic Rocks. In JOHNSON, G.A.L. (Ed) Robson's Geology of North East England. *Trans. Nat. Hist. Soc. Northumb.,* **56**, 283-295.

SMITH, D.B. & FRANCIS, E.A. 1967. Geology of the country between Durham and West Hartlepool. *Mem. Geol. Surv. G.B.,* Sheet 27, 354pp.

SUMMERS, T.P., LOS, A.P. & WESTBROOK, G.K. 1982. Geophysical investigations of the Whin Sill in the Roman Wall District of Northumberland. *Proc. Yorkshire Geol. Soc.,* **44**, 109-118.

TATE, G. 1867. The geology of the district traversed by the Roman Wall. Appendix to BRUCE, J.C. *The Roman Wall.* 3rd edition. London: Longmans, 441-454.

TAYLOR, B.J., BURGESS, I.C., LAND, D.H., MILLS, D.A.C., SMITH, S.B. & WARREN, P.T. 1971. *British regional geology: northern England.* 4th edition. London: HMSO, 121pp.

TEALL, J.J.H. 1884. Petrological notes on some north of England dykes. *Q.Jl geol. Soc. Lond.,* **40**, 209-247.

Geology of Hadrian's Wall

TROTTER, F.M. 1929. The Glaciation of Eastern Edenside, the Alston Block and the Carlisle Plain. *Q. Jl geol. Soc. Lond.*, **85,** 558-612.

TROTTER, F.M. & HOLLINGWORTH, S.E. 1932. The Geology of the Brampton District. *Mem. Geol. Surv. G.B.* Sheet 18, 223pp.

WALKER, D 1966. The Late Quaternary history of the Cumberland lowland. *Phil. Trans. R. Soc.,* **B 251,** 1-210.